Mohamed Abdel-Raheem
Mai Magdy Mousa

Entomopathogenic Fungi and its Biological Control

Mohamed Abdel-Raheem
Mai Magdy Mousa

Entomopathogenic Fungi and its Biological Control

Noor Publishing

Imprint

Cover image: www.ingimage.com

Publisher:
Noor Publishing
is a trademark of
Dodo Books Indian Ocean Ltd. and OmniScriptum S.R.L publishing group

120 High Road, East Finchley, London, N2 9ED, United Kingdom
Str. Armeneasca 28/1, office 1, Chisinau MD-2012, Republic of Moldova, Europe
Managing Directors: Ieva Konstantinova, Victoria Ursu
info@omniscriptum.com

Printed at: see last page
ISBN: 978-620-8-87052-2

Entomopathogenic Fungi and its Biological Control

By

Prof. Dr. Mohamed Abdel-Raheem

Eng. Mai Magdy Mousa

(2025)

Entomopathogenic Fungi and its Biological Control

Prof. Dr. Mohamed Abdel-Raheem Ali Abdel-Raheem

Professor of Entomology (Biological Control)

Pests & Plant Protection Department

Agricultural and Biological Research Institute

National Research Centre.

Eng. Mai Magdy Mousa

Shebin El-Kom, Menoufia, Egypt

Email: Maimagdy1427@gmail.com
Mobile: (+2) 0 121 123 0330

Content list

Introduction

Plant diseases caused by fungi are included in the category of agricultura l pests. It is estimated that there are approximately 67,000 different species of pests . These pests pose a constraint on agricultural production and are responsible for ca using around 40% of potential global crop yield losses. Despite extensive efforts in pest control, these losses persist. This suggests that focusing on pest management i s the best approach to improve yields and ensure access to food. Pest problems are an inevitable part of agriculture, mainly because agricultural systems are simplified and not stable modifications of natural ecosystems.

The disruption of ecological forces that regulate potential insect pest spe cies in natural ecosystems is caused by agricultural land. Farming practices can cre ate a more favorable environment for pest activity, such as through irrigation or wa rm conditions in glasshouses. Additionally, the use of pesticides can have detrimen tal effects on the natural predators and parasites that help keep pests under control. One of the major issues arises when pests develop resistance to insect pesticides. C urrently, over 500 arthropod species have shown resistance to one or more types of insect pesticides.

Another significant problem is posed by alien species that are unintentio nally introduced to new countries or continents and lack evolved natural predators. These alien species cause the most damage to crops. Emerging pests also pose a thr eat. Farmers face pressure to reduce the use of insect pesticides without compromis ing yields or crop quality. However, controlling pests is becoming increasingly cha llenging due to pesticide resistance and the decreasing availability of products. Bio logical control, which involves using living organisms to manage the population de nsity of pest organisms, is a strategy employed by humans. Natural enemies, which are organisms that kill or weaken other organisms, are used in biological control p rograms worldwide, particularly in agriculture.

Biological control agents encompass the following categories:

(1) Predatory insects and mites, which consume their prey.

(2) Parasitoids, which are insects with a freeliving adult stage and a larval stage tha t parasitizes another insect.

(3) Parasites and microbial pathogens, such as nematodes, fungi, bacteria, viruses, and protozoa, which cause lethal infections.

In Europe, these agents are employed in various ways, for instance:

(1) The application of parasitoids in glasshouses to control whiteflies.

(2) The use of parasitic nematodes against slugs.

(3) The deployment of mycoparasitic fungi to manage plant diseases in horticultur al crops.

(4) The utilization of viruses to regulate codling moth populations in apple orchar ds, as well as the creation of habitats on farms to increase the numbers of natur al predators and other beneficial organisms. Natural enemies constitute a signif icant portion of the world's biodiversity; for example, parasitoids account for a pproximately 10% of all species worldwide, and each agricultural pest species may be predated upon by hundreds of different natural enemies. However, onl y a small fraction of the available species have been studied for their potential role in crop protection.

As a result, even though there are 750 known species of insect pathogeni c fungi, only a meager number of them, less than 20, have been seriously consi dered as control agents for insect pests. The natural enemies present in agro ec osystems play a crucial role in preventing pests from reaching harmful levels. The methods employed in utilizing biological control agents vary depending o n the type of pest, the biological characteristics of the control agent, and the ag ricultural context. Individuals utilizing biocontrol are obligated to adhere to the FAO's code of conduct on the import and release of biological control agents, as well as national regulations pertaining to releases into the environment. The use of microbial agents is regulated by Plant Protection Product regulations, w herein authorizations follow a similar course to those for chemical pesticides, a lthough there are also specific schemes in place. Biological control agents poss ess numerous desirable attributes, including host specificity, absence of toxic r esidue, lack of phytotoxic effects, human safety, and the potential for self-sustaining pest management. Additionally, many of these agents possess the ab ility to actively locate their prey.

Producing BCAs locally can be crucial in selecting the appropriate natural enemies for smallscale needs. The successful utilization of BCAs necessitates a deep under standing of both the ecological dynamics of the natural enemy and the pest. When t hese conditions are met and the use of BCAs is integrated into IPM practices, biolo gical control can sometimes be more costeffective compared to chemical control al one. However, in the current market system, many biological control products have struggled to compete against cheaper and more effective synthetic pesticides. The drawbacks of BCAs include their limited applicability, delayed effectiveness in pes t control, potential lack of persistence in the environment, and unpredictable effica cy, especially in outdoor settings. The approval process for microbial agents also in curs significantly higher authorization costs compared to macro-agents.

Nowadays, farmers and growers are well acquainted with the use of predators and parasitoids to control arthropod pests. However, there is also the option of utili zing specific microorganisms that can eliminate arthropods. These microorganisms , known as entomopathogens, encompass fungi, nematodes, bacteria, and viruses, which are naturally prevalent and can cause infections in various pest species.

Entomopathogens play a role in regulating the populations of arthropods in a natural manner. A significant amount of research in this field focuses on the causal agents of insect diseases and their exploitation for biological pest control. Many en tomopathogens can be produced on a large scale, formulated, and applied to pest p opulations in a way similar to chemical pesticides, namely as nonpersistent treatme nts that are released in large quantities.

Additionally, entomopathogens have been utilized as classical biological cont rol agents for nonnative insect pests, and their effectiveness has been further enhan ced through manipulation of their habitat. Entomopathogenic fungi, for instance, ar e believed to consist of approximately 750 species that can infect insects or mites. While these fungi can attack a wide range of hosts, each specific species and strain is highly specific in its target.

The fungi release spores that infect their host by germinating on its surface an d subsequently infiltrating its body. The host typically dies within a period of 4 to 10 days, depending on the fungus type and the number of spores infecting it. After death, the fungi generate numerous new spores on the deceased body, which then d isperse and continue their life cycle on new hosts. Two primary taxonomic orders o f entomopathogenic fungi exist.

The first order, known as Entomophthorales, belongs to the phylum Zygomyc ota and includes genera like Pandora, Entomophthora, and Conidiobolus. Many of these species are coevolved, obligate pathogens that exhibit ecomorphological ada ptations tailored to the life cycles of their hosts, such as the production of forcefull y ejected infective spores that are released from insect cadavers during nighttime w hen environmental conditions are most favorable for infection.

"These fungi frequently cause natural outbreaks in populations of insects and mites. However, some of them pose a challenge for mass production in a controlled environment, making it difficult for researchers to develop them as biopesticides. Nonetheless, they can still be utilized for pest control by applying them to the field in the form of mycelium or infected insect cadavers. For instance, isolates of Neoz ygites floridana from South America are being investigated for the biological contr ol of the cassava green mite Mononychellus tanajoa, which is a significant invasive pest of cassava in Africa.

In other cases, spontaneous outbreaks of entomophthoralean fungi are being h arnessed as a natural method of pest control. For example, a prediction service crea ted by Don Steinkraus at the University of Arkansas is employed to inform cotton farmers in the southeastern United States about occurrences of Neozygites fresenii in cotton aphids, Aphis gossypii. These outbreaks have the potential to decrease ap hid populations by roughly 80% in just 4 days and can even be forecasted up to ten days in advance. Accordingly, cotton farmers have the option to abstain from expe nsive insecticide sprays.

The second prominent category of entomopathogenic fungi, called the Hypocr eales, can be found within the Ascomycotina phylum. Within the Ascomycotina, th ere exist numerous species that only reproduce asexually and lack knowledge of th eir sexual phase (teleomorph), with confirmation of their classification as members of the Ascomycota solely relying on molecular (DNA) data. These fungi are com monly referred to as anamorphic fungi. On the other hand, sexually reproducing hy pocrealean fungi can be found in the genera Cordyceps and Torrubiella. Cordyceps has its origins in Asia and its species are important natural parasites of various ins ects and mites found in tropical forests. Cordyceps sinensis, for instance, is a patho gen of ghost moth (Thitarodes) caterpillars that reside at high altitudes in the Himal ayas, Tibet, and China.

In China, it is known as dong chong xia cao, or "winter animal summer grass. " It holds significant value as an herbal medicine and contains numerous pharmalo gically active compounds. Notable anamorphic genera of entomopathogenic fungi i nclude Beauveria, Isaria, Metarhizium, and Lecanicillium. Species from all of thes e genera are utilized as biopesticides for insect pests. Recently, molecular methods have aided in uncovering the teleomorph connections of these fungi. Moreover, the se molecular studies have shed light on the phylogenetic relationships between thes e fungi and other species. This research is being spearheaded by prominent mycolo gists in the United States, such as Meredith Blackwell, Richard Humber, Steve Reh ner, and Joey Spatafora."

It has been demonstrated that the ability to infect insects has evolved multiple time s in the ascomycete fungi, with a shift in host preference from plants to insects and vice versa. Many species of the nonsexual entomopathogenic fungi can be easily c ultivated on various media, making them suitable for use as biopesticides in the ma ss treatment of pest populations. These fungi possess several desirable traits, including safety for humans, compatibility with other natural enemies, and the abse nce of toxic residues. Additionally, they have the potential to provide longlasting control by multiplying within the pest population. Due to their contact action, they are particularly effective against sapsucking pests like aphids and whiteflies, which cannot be infected by other types of biopesticides that require ingestion, such as ba cteria and viruses. However, their higher cost compared to chemicals and susceptib ility to certain environmental conditions are disadvantages.

There is also a misconception that biopesticides will have the same effectiveness as chemical insecticides, often leading to the overlooking of their desirable biological characteristics. Furthermore, the high registration costs relative to the current mark et size discourage companies from seeking registration. Currently, there are two co mmercial products available for UK growers, namely Vertalec (used for controllin g glasshouse aphids) and Mycotal (used for controlling glasshouse whitefly). Both products are based on the fungus species Lecanicillium: Vertalec relies on Lecanici llium longisporum, while Mycotal is based on Lecanicillium muscarium (previousl y thought to be Verticillium lecanii, but the classification has since been revised). These products are manufactured by Koppert BV and were originally developed at the Glasshouse Crops Research Institute, a predecessor of Warwick HRI. It is expe cted that products based on Beauveria bassiana and Metarhizium anisopliae will so on be available in the UK for controlling glasshouse pests and vine weevils.

Biological control is an essential aspect of a comprehensive approach to mana ging pests, aiming at minimizing their populations through the use of Pathogens. Pathogens encompass various diseasecausing organisms such as bacteria, fungi, and viruses.These microorganisms incapacitate or eliminate their host and usually target specific insect groups.Infestations of flies infected with fungi are frequently observed after periods of elevated humidity.

Biological-based pesticides

Microorganisms such as viruses, fungi, protozoa, and bacteria are the primary components of this type of pesticide. Each microorganism contains a specific activ e ingredient that targets particular pests, but microbial pesticides can effectively co ntrol or eradicate various pests that harm crop production. One example is the use of fungi to control weeds or insects that are harmful to crops.

Characteristics of biopesticides

Biopesticides possess certain characteristics:

1. Biopesticides have a narrow target range and are used for specific pest issues.
2. Biopesticides work at a slower pace compared to other pesticides.
3. Rather than completely eliminating pests, biopesticides aim to suppress their population.
4. The timing of biopesticide application is crucial.
5. Biopesticides have limited field persistence and shelf life.
6. Biopesticides are often used in integrated pest management (IPM) programs.
7. They are more environmentally friendly and safer for humans compared to conventional pesticides.
8. Biopesticides do not typically leave behind residue issues.

"Utilization of biopesticides to enhance the safety of food products."

Food safety necessitates the extensive use of biopesticides, which encompass various types such as beneficial insects, microbial pesticides, microbial toxins, and antibiotics. Beneficial insects refer to those that feed on pests, including mites and spiders in the case of arthropods. Microbial pesticides can be naturally occurring or genetically modified fungi, protozoans, bacteria, algae, or viruses, offering an effe ctive alternative to chemical insecticides. Microbial toxins, derived from microorga nisms like fungi or bacteria, have the capability to cause death or rupture the gut of pests due to their high toxicity. Studies have demonstrated the significant role of i nsecticidal toxins in the pathogenesis of these entomopathogens. Antibiotics, whic h can kill or inhibit the growth of microorganisms such as fungi and bacteria, play a crucial role in crop protection. The success of biotechnology has resulted in the d evelopment and commercialization of transgenic crops resistant to major pests. Firstgeneration transgenic crops contain single insecticidal Bt genes, providing resi stance against cotton and corn pests. There is growing interest in exploring plants a nd their chemobiodiversity as a potential source of bioactive substances for crop pr otection. Plants have the ability to synthesize a vast array of secondary metabolites, which are small organic molecules with complex and unique carbon skeleton struc tures. These substances have been used for many years to protect crops and can be released into the environment when plants are utilized as smother crops, green man ures, intercrops, cover crops, mulch, or in rotational sequences, resulting in effectiv e management of insect pests and pathogenic diseases, leading to improved crop yi elds. Botanicals encompass both crude or semirefined extracts and isolated or purif ied compounds from various plant species, which are also available as commercial products.

Biopesticides derived from microorganisms

Entomopathogenic fungi play a crucial role in regulating the population of var ious insects in nature. Alongside agricultural insect pests, these fungi have the pote ntial to be effective agents in the control of insects.

These fungi primarily infect their host by penetrating the cuticle, entering the host's hemolymph, producing toxins, and consuming nutrients present in the hemocoel to avoid triggering immune responses in insects.

There are a total of 1600 different types of viruses that infect 1100 different specie s of insects and mites. Baculovirus is a distinct group of viruses that specialize in i nfecting about 100 species of insects.

Bacterial biopesticides are widely used and costeffective for pest control. They are effective against various species of butterflies, moths, flies, beetles, and mosquitoe s. The bacteria used as pathogens are mostly spore-forming and rod-shaped bacteria belonging to the Bacillus genus.

Entomopathogenic nematodes (EPNs) introduce symbiotic and pathogenic bacteria such as Xenorhabdus (in the Steinernematidae) or Photorhabdus (in the Heterorha bditidae) into the hemocoel of their hosts. The bacteria multiply and eventually lea d to the death of the host, which can occur within 48 hours of infection.

Infected insects exhibit abnormal behavior compared to uninfected individuals. Wh ile the nematode alone can kill its host, it cannot reproduce without the associated bacterium.

These nematodes have also been successfully used in controlling citrus root weevils, turf pests, and in mushroom crops .

The researchers conducted a randomized controlled trial involving 200 partici pants with heart disease. The participants were randomly assigned to two groups: t he treatment group and the control group. The treatment group received the new dr ug while the control group received a placebo. The researchers measured various p arameters such as blood pressure, heart rate, and cholesterol levels before and after the treatment. They found that the new drug significantly reduced blood pressure a nd heart rate compared to the placebo. The drug also showed a positive effect on lo wering cholesterol levels in the treatment group. These findings suggest that the ne w drug could be an effective treatment option for heart disease. Further research is needed to validate these results and assess the long-term effects of the drug.

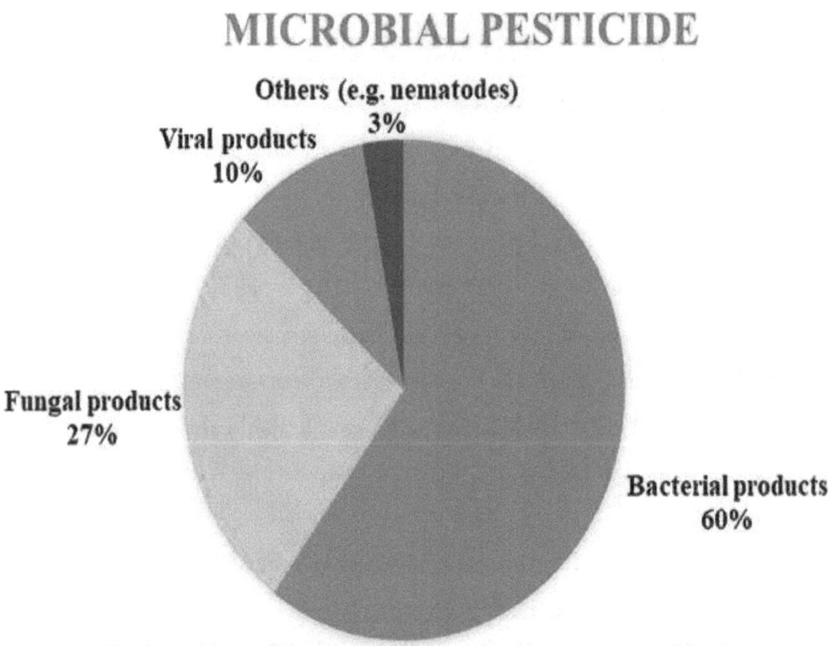

Microbial biopesticide products worldwide

The active ingredient in microbial biopesticide products used globally is microorga nisms such as bacteria, fungi, viruses, and nematodes, as stated by the Division of Agricultural Chemicals in 2011.

Groups of entomopathogens

Below, we provide a description of significant groups of entomopathogens and thei r methods of infection.

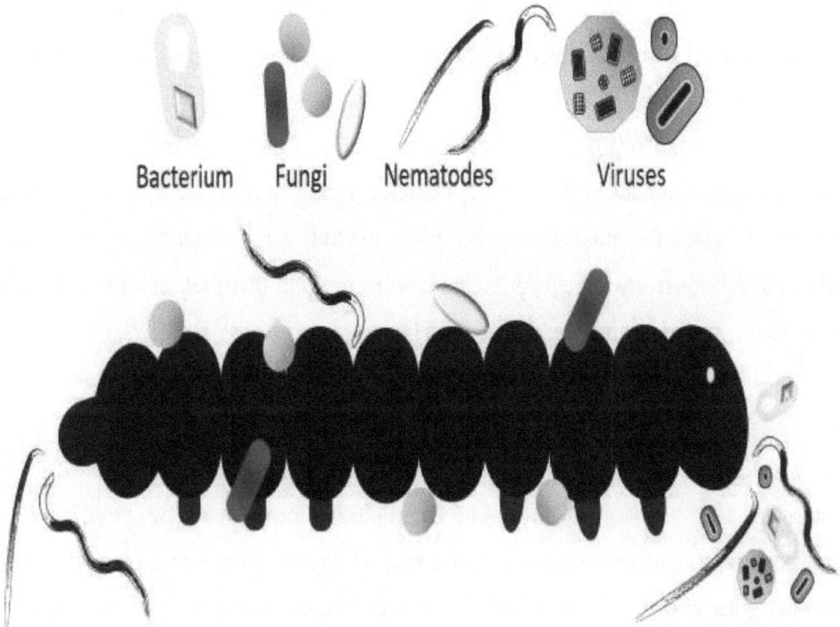

Infection routes of entomopathogens

Entomopathogenic fungi have played a significant role in the history of controlling insects through microbial means. Evidence from history suggests that entomopathogenic fungi were the first to be identified as causing disease in insects [1].

One type of entomopathogenic fungus is Beauveria bassiana, also known as white muscardine fungus, which can infect both immature and adult insects. Another type is Metarhizium anisopliae, referred to as green muscardine fungus, which has the ability to infect 200 species of insects and arthropods. These fungi are found in soil and are widely distributed. They have been observed to naturally occur in over 750 species of insects [2-4].

The natural environment for these fungi is soil, as they deposit their infectious spores there and remain in the soil for a certain period of their life cycle. Therefore, soil is considered the most appropriate place to study their presence [5]. Previous studies have extensively investigated the occurrence and distribution of insect pathogenic fungi in agricultural field soils [6-11].

Additionally, there have been studies on isolating these fungi from insect cadavers [12-17].

Wild isolates of these fungi are still highly valuable due to their potential unique characteristics in the biological control of insect pests. The presence of specific entomopathogenic fungal species can indicate their ability to survive in a particular environment. This information is useful for selecting biocontrol agents, as the dominat species in a given location are typically the most suitable candidates [18].

Factors such as geographical location, climate, habitat, altitude, and soil pH can influence the presence of fungal species, and each species may respond differently to these conditions [18]. The Galleriabait technique is commonly used to isolate entomopathogenic fungi from soil [19-22].

While identifying common entomopathogenic fungi like Beauveria bassiana, Metar hizium anisopliae, and Verticillium lecanii may seem simple even to those without expertise in mycology, it is important to note that these species also possess compl ex molecular, morphological, and pathobiological traits [23, 24].

Entomopathogenic fungi mainly cause infections when their spores come into contact with the arthropod host. Under favorable conditions of moderate temperatu res and high humidity, fungal spores germinate and penetrate the insect's cuticle thr ough enzymatic degradation and mechanical pressure, gaining entry into the insect' s body. Once inside, the fungi multiply, invade the insect's tissues, emerge from the dead insect, and produce more spores. Natural outbreaks of entomophthoralean fu ngi, such as Entomophaga maimaiga, Entomophthora muscae, Neozygites fresenii, N. floridana, and Pandora neoaphidis, have been known to significantly reduce hos t populations. Despite being difficult to culture in artificial media and not suitable f or commercial sale as biopesticides, these fastidious fungi still play a crucial role i n the natural control of certain pest species.

On the other hand, hypoclealean fungi such as Beauveria bassiana, Isaria fum osorosea, Hirsutella thompsonii, Lecanicillium lecanii, Metarhizium acridum, M. a nisopliae, and M. brunneum are commercially available as biopesticides in various formulations worldwide. These fungal pathogens have a wide range of hosts and ar e especially effective against pests with piercing and sucking mouthparts, as the sp ores do not need to be ingested. However, entomopathogenic fungi are also effecti ve against pests with chewing mouthparts like wireworms and borers.

In relation to fungi, the sporeforming microsporidium Paranosema (Nosema) locustae is a pathogen that has been used to control locusts, grasshoppers, and certa in crickets. When ingested, this pathogen infects the midgut tissues and subsequent ly spreads to the fat body tissues. The disease weakens the orthopteran host and ev entually leads to its death within a few weeks.

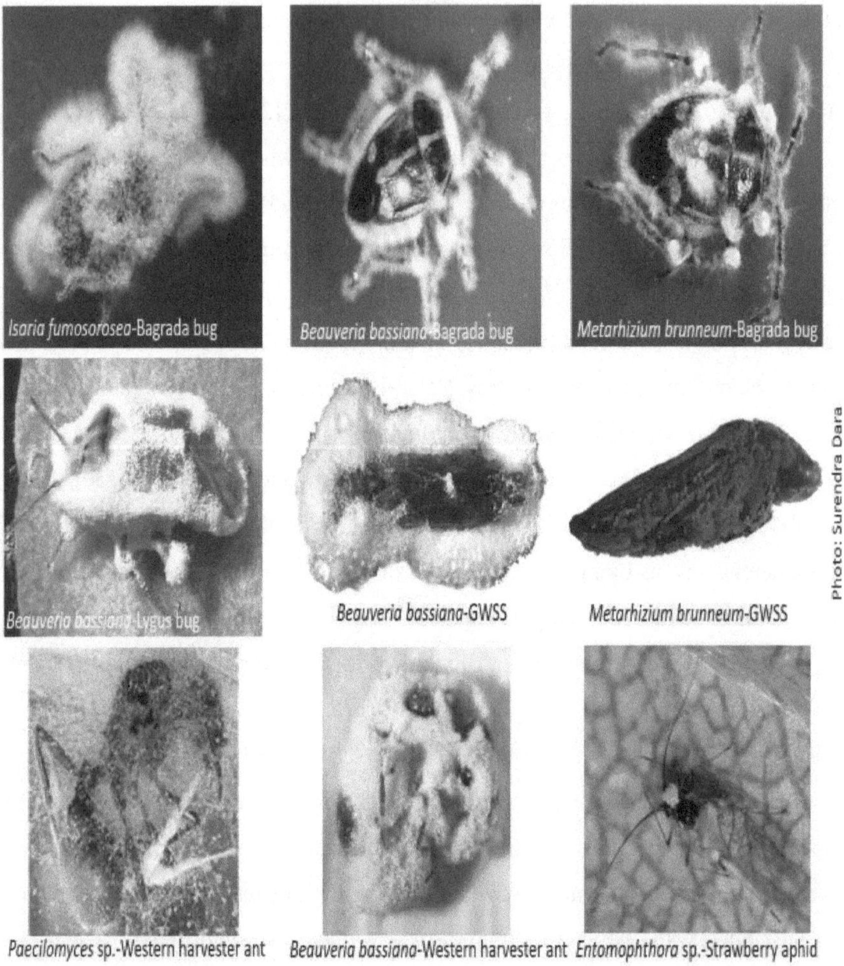

Fig. (1): "Insects that are contaminated with Entomopathogenic Fungi"

There are certain types of fungi that have the ability to invade and eliminate insects . These fungi are referred to as entomopathogenic fungi or insect pathogenic fungi. They specifically target insects and usually have a preference for particular insect s pecies, making them harmless to animals and plants. While most of these fungi req uire moisture for infection and growth, there are some species that can infect insect s even in dry conditions. Mother Nature offers her assistance free of charge!

The fungus emerges from the body of an insect that has been killed by it, resul ting in the appearance of spores that can be observed as a dusty or cottony appeara nce.

If you come across numerous insects that are dead and covered in fungus, you can consider creating your own bio-insecticide using these dead insects.

Gather as many dead insects as possible from the field, place them in a jar wit h water, lightly crush them, and stir thoroughly.

This process will release the spores of the fungus into the water. Filter the wat er slightly to remove larger parts of the insects. The remaining solution can be utili zed for testing.

Spray the solution onto insects placed in a jar and observe if these insects bec ome infected in the following days. For comparison, spray another jar of insects wi th pure water as a control.

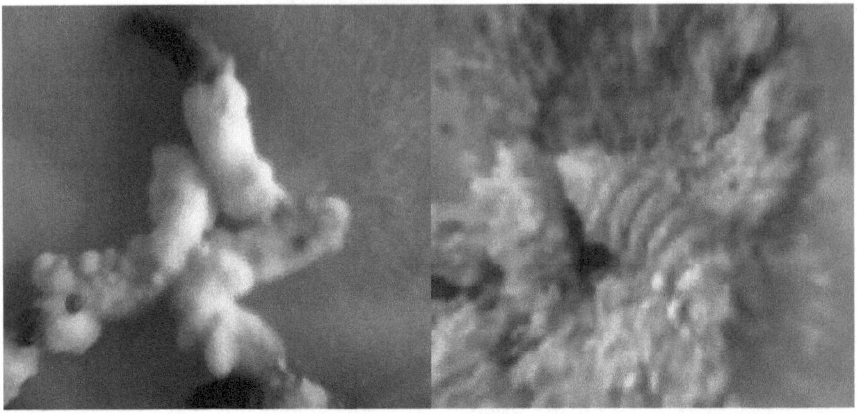

Beauveria bassiana *Metarhizium anisopliae*

Fig. (2): Insects that have been invaded by Entomopathogenic Fungi

The mode of action of entomopathogenic fungi

Can be described in nine steps, as outlined by Roberts

These steps include the attachment of infective units (such as conidia or zoo spores) to the insect's outer layer, the germination of these units on the insect's cuti cle, and the penetration of the cuticle either through germ tubes or infection pegs fr om appressoria.

After penetrating the cuticle, the fungi multiply in the haemocoel in the yeast phase , followed by the production of toxic metabolites. This eventually leads to the deat h of the host organism.

Subsequently, the fungi continue to grow in the mycelial phase, invading virtually all host organs. They penetrate the insect's interior and reach the exterior through t he cuticle, and finally, new infective units (conidia) are produced.

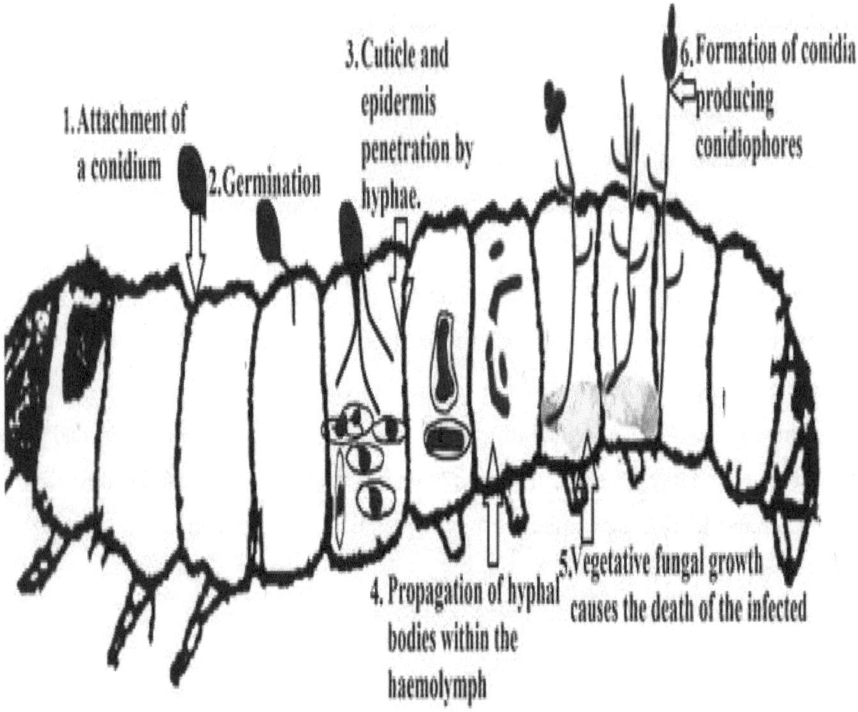

Fig. (3): The mechanism by which Entomopathogenic Fungi exert their effects

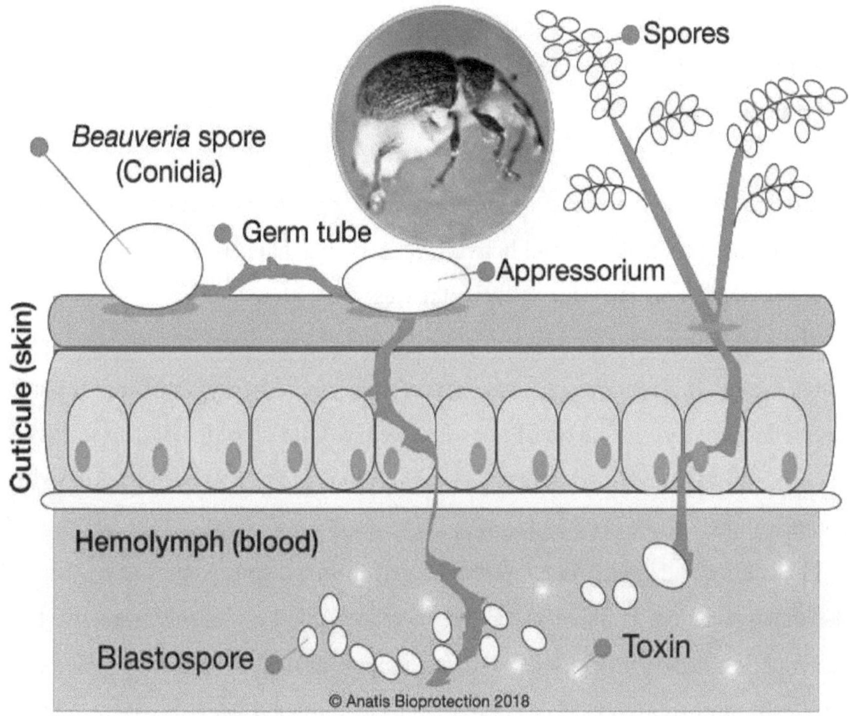

Fig. (4): Mechanism employed by Entomopathogenic Fungi

Separation of Entomopathogenic fungi

Soil isolation

Soil samples were collected every two weeks from three fruit orchards in the area s urrounding Alexandria desert road between January and June 2015. A total of 100 soil samples were collected during the entire season.

These orchards consisted of a mixture of date palms, olives, mangoes, annonas (keshta fruit), pomegranates, pears, grapes, and other types of fruit trees. T he orchards were a combination of trees that were at least 7 years old. The soil in t hese orchards was sandy and contained organic material that had been transferred f rom cattle barns.

The entomopathogenic fungi were retrieved from the soil samples using the Galleriabait technique, as described in the reference [25]. This technique involved collecting soil samples from a 1meter diameter circle around the tree trunks, at a de pth of 0t020 cm. Approximately 1.5 kg of soil was collected in plastic bags and tra nsported to the laboratory in a cooled box. In the laboratory, the soil samples were divided into three parts and placed in plastic containers. Larvae of the greater wax moth, Galleria mellonella, were added to the containers at a rate of 3 larvae per cup and moistened to about 15% water content.

After incubating the samples at 25°C for 7 days, any infected larvae were tran sferred to 5 cm diameter Petri dishes lined with wet filter paper and kept for further development and sporulation. The number of positive samples (samples containin g entomopathogenic fungi) and the number of infected G. mellonella larvae in each sample were recorded, as mentioned in [26].

The fungi that were extracted were identified based on their morphological charact eristics of reproductive structures, following the guidelines of [26, 27]. The isolates that were obtained were identified as Beauveria bassiana, Metarhizium anisopliae, and Verticillium lecanii. The natural abundance of these entomopathogenic fungi i n the studied fruit orchards is shown in Table 1, indicating the percentages of positi ve samples and percentages of infected G. mellonella larvae for each sampling date . Among the collected samples, the most abundant fungus was B. bassiana, account ing for 25.03% of the total samples, followed by M. anisopliae at 17.76%, and V. l ecanii at 14.49%. B. bassiana was present from January to April, with the highest o ccurrence in early January and the lowest in mid-April.

The occurrence of this fungus in positive samples ranged from 10 to 70% for each sampling date. The percentages of infected G. mellonella larvae in the collect ed samples varied from 1.11 to 17.78%. Our findings demonstrated a high prevalen ce of entomopathogenic fungi in the soil samples collected from fruit orchards. A s urvey conducted by [26] found that positive soil samples containing B. bassiana or M. anisopliae accounted for less than 2% of the total samples collected from differ ent regions in Egypt. Similar to our study, *B. bassiana* was the most frequently fou nd entomopathogenic fungus in the soils, followed by *M. anisopliae* in Spain [25]. In China, *B. bassiana* was more prevalent than *M. anisopliae* in the soils [26].

Table 1:

The percentage of soil samples from fruit orchards along Alexandria desert road th at tested positive for the entomopathogenic fungi *Beauveria bassiana, Metarhizium anisopliae*, and *Verticillium lecanii* varied depending on the date of sampling, acc ording to a study [26].

	Beauveria bassiana		*Metarhizium anisopliae*		*Verticillium lecanii*	
Sampling dates	Positive samples	Infected larvae	Positive samples	Infected larvae	Positive samples	Infected larvae
1.1.15	70	14.44	10	4.44	0.0	0.0
15.1.15	40	17.78	50	6.67	30	12.22
1.2.15	60	16.67	20	2.22	10	3.33
15.2.15	30	3.33	10	1.11	10	1.11
1.3.15	20	2.22	0.0	0.0	40	5.56
15.3.15	10	1.11	0.0	0.0	0.0	0.0
1.4.15	0.0	0.00	0.0	0.0	0.0	0.0
15.4.15	30	3.33	40	4.44	0.0	0.0
1.5.15	0.0	0.0	10	3.33	0.0	0.0
1.6.15	0.0	0.0	0.0	0.0	0.0	0.0
Mean	25.03	7.33	17.76	2.40	14.49	3.99

***M. anisopliae* was present during two time periods.**

The first period was from early January to early March, and the second period was from mid April to early May. During the second period, there was a peak of 5 0% positive samples and a minimum of 10%. The percentage of G. mellonella larv ae infected with *M. anisopliae* in the collected soil samples ranged from 1.1% to 6. 67%. *V. lecanii* was found from mid January to early March, with 10% to 40% pos itive samples and 1% to 12.2% infected G. mellonella larvae. This information is s upported by reference [43].

Fig 1 displays the percentages of soil samples containing various entomopath ogenic fungi. B. bassiana was detected in the soil beneath all studied fruit trees, acc ounting for 10 to 40% of the samples collected. This fungus was most commonly found under Mang o trees (50% of samples) and Pomegranates (40%). Its occurrence ranged from 10-30% for the other types of fruit. M. anisopliae was discovered under seven differen t fruit types, with a fairly even distribution among them. Its occurrence ranged fro m 10-30 % of the collected samples. V. lecanii was found under six different fruit types, but there was no clear relationship observed between the fungus type and the fruit kind [26].

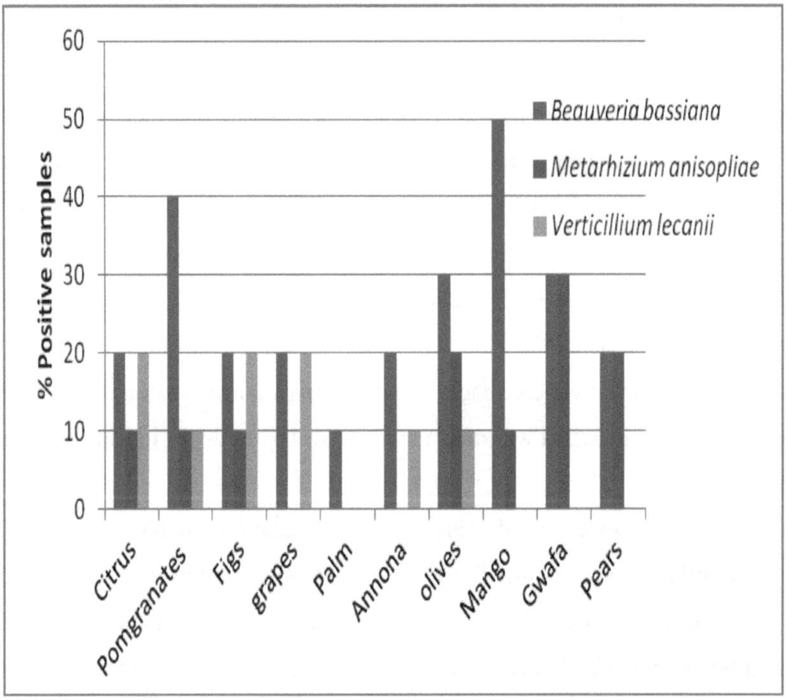

Fig 5:

The presence of *Beauveria bassiana*, *Metarhizium anisopliae*, and *Verticillium lecanii* is plentiful in orchards near Alexandria desert road, depending on the type of fruit [26].

It was not possible to establish a definite correlation between the occurrence of entomopathogenic fungi and the specific type of fruit tree. Fungus occurrence was relatively higher (60%) in olives, mangos, pomegranates, and guava as compared to other types of fruits. Palms, on the other hand, had a lower occurrence rate of only 10% for fungi.

Fungal distribution among fruit trees is influenced by both the specific fungus species and the rate of organic fertilization applied beneath the trees rather than the type of tree itself. Typically, these particular trees are given greater amounts of manure compared to others. In Egypt, it is uncommon for farmers to fertilize palm trees [25]. According to Charnley [26], factors such as organic content and temperature play a role in determining the abundance and activity of fungi in the soil.

Isolation of Insect Corpses

The insect corpses were gathered from the field in Kafr El-Sheikh Governorate and placed in Petri-dishes measuring 9 cm in diameter and 1.5 cm in height at a temperature of 24 °C. These dishes were examined on a daily basis to monitor the development of fungi, which were later purified and utilized to verify the disease cycle (according to Koh's postulates). After that, the fungi were stored on slants of PDA artificial media at a temperature of 4 °C until they were employed in future experiments [12].

New fungal generations from infected insects were obtained through a process of isolation. The insects, which were covered in fungal mycelium, were collected from the field and brought to the laboratory. It was observed that various stages of Cassida vittata and Scrobipalpa ocellatella had died and were completely covered with fungal mycelium. These insect cadavers were then placed on a wetted filter paper in a Petri dish and incubated at a temperature of 24 ± 1 ° c for a period of seven days. From the surface of the insect cadavers, a new fungal generation was isolated and subsequently cultured on PDA medium in Petri dishes.

The fungal cultures were purified on a weekly basis until pure cultures were obtained.

As stated in table (2), two isolates of Beauveria bassiana and Metarhizium anisopliae were derived from infected cadavers of Cassida vittata and Scrobipalpa ocellatella found in two regions, Beila and Abu Kalab, within Kafr ElSheikh Governorate [12].

Table (2):

Between February 2000 and May 2003, entomopathogenic fungi were discovered i n insect cadavers from various locations in Kafr El-Sheikh[12].

Fungi	Original host	Host plant	Location	Date
B. bassiana	*S. ocellatella* (4 th larvae)	Sugar -beet	Abu kalab	May 2001
B. Bassiana	*C. vittata* (Larvae-adults)	Sugar- beet	Abu Kalab	April 2001
M. anisopliae	*C. vittata* (Larvae-adults)	Sugar- beet	Beila	April 2003

Table (3):

The incidence rates of positive entomopathogenic fungi recovered by trapping *G. mellonella* larvae from soil samples collected from three governorates are docu mented [26].

Locality	Number of samples	Fungi positive samples and incidence (%)	
		B. bassiana	*M. anisopliae*
Elbehaira	119	3 (2.52%)	1 (0.84%)
Kafr El-Shaeikh	103	2 (1.94%)	2 (1.94%)
Aswan	97	1 (1.03%)	1 (1.03%)
Total	319	6 (1.88%)	4 (1.25%)

Entomopathogenic fungi were cultivated and maintained using two different methods:

1. Peptone medium: This medium consisted of 10g of peptone, 40g of dextrose, 2g of yeast extract, 15g of agar, and 500 ml of chloramphenicol, brought to a total of o ne liter with distilled water. The medium was sterilized at 120oC for 20 minutes an d poured into Petri dishes (9 cm diameter x 1.5 cm). The fungi were then inoculate d onto the medium and incubated at a temperature of 25 ±2oC with a relative humi dity of 85 ±5. The fungal isolates were recultured every 15-
30 days and stored at 4oC.

2. Potato dextrose agar (PDA): This medium was prepared by mixing 250g of potatoes, 20g of agar, and 1000 ml of distilled water. After autoclaving the mediu m at 120°C for 20 minutes, it was poured into Petri dishes (9 cm diameter x 1.5 cm).
The fungal isolates were recultured every 14-
30 days and stored at 4oC. To restore the virulence of the isolates, they were either passed through their natural host or through Galleria mellonella larvae, which are wax moth larvae.

In the case of *B. bassiana* and *M. anisopliae* isolates, wetted rice was used for prop agation. Two kilograms of wetted rice were washed in boiled water for 10 minutes and placed in thermal bags. These bags were then autoclaved at 120°C for 20 minut es and infected with the isolates. The bags were incubated at 25 ±2°C for 15 days. The conidia were collected using distilled water and filtered through cheese cloth to remove mycelium clumps. Tween 80% was added to the harvested conidia, as described by Lacey [27].

Here are a few examples of the most frequently encountered fungi that can infect insects:

Beauveria bassiana is a type of fungus.

Beauveria bassiana is a naturally occurring fungus found in soils worldwide.

It parasitizes various arthropod species, causing a disease known as white muscardine, making it one of the entomopathogenic fungi.

This fungus is utilized as a biological insecticide to manage numerous pests, including termites, thrips, whiteflies, aphids, and various types of beetles.

Preferred Scientific Name

- *Beauveria bassiana* (Bals.-Criv.) Vuill. (1912)

Preferred Common Name

- white muscardine fungus

Other Scientific Names

- *Beauveria brongniartii* subsp. delacroixii Sacc. (1892)
- *Beauveria densa* (Link) F. Picard (1914)
- *Beauveria doryphorae* R. Poiss. & Patay (1935)
- *Beauveria effusa* (Beauverie) Vuill. (1912)
- *Beauveria laxa* Petch (1931)
- *Beauveria paradoxa* (Bals.-Criv.) Ramsb. (1941)
- *Beauveria shiotae* (Kuru) Langeron (1936)
- *Beauveria stephanoderis* (Bally) Petch (1924)
- *Beauveria sulfurescens* (van Beyma) J.J. Taylor (1970)
- *Beauveria tenella* (Sacc.) MacLeod (1954)
- *Beauveria vexans* (Pettit) Petch (1926)

- *Beaveria delacroixii* (Sacc.) Petch (1924)
- *Botrytis bassiana* Bals.-Criv. (1835)
- *Botrytis bassiana* subsp. tenella (Sacc.) Sacc. (1882)
- *Botrytis bassiana* var. lunzinensis Szilvinyi (1941)
- *Botrytis delacroixii* (Sacc.) Delacr. (1893)
- *Botrytis effusa* Beauverie (1911)
- *Botrytis necans* Massee (1914)
- *Botrytis paradoxa* Bals.-Criv. (1835)
- *Botrytis stephanoderis* Bally (1923)
- *Botrytis tenella* Sacc. (1881)
- *Isaria citrinula* Speg. (1910)
- *Isaria densa* (Link) Giard (1891)
- *Isaria shiotae* Kuru (1931)
- *Isaria tenella* (Sacc.) Giard (1891)
- *Isaria vexans* Pettit (1895)
- *Penicillium bassianum* (Bals.-Criv.) Biourge, 1923
- *Penicillium delacroixii* (Sacc.) Biourge (1923)
- *Penicillium densum* (Link) Biourge (1923)
- *Spicaria bassiana* (Bals.-Criv.) Vuill. (1910)
- *Spicaria delacroixii* (Sacc.) Vuill. (1910)
- *Spicaria densa* (Link) Vuill. (1910)
- *Sporotrichum densum* Link (1809)
- *Sporotrichum globuliferum* Speg. (1880)
- *Sporotrichum larvatum* Peck, 1879
- *Sporotrichum larvicola* Peck (1887)
- *Sporotrichum minimum* Speg. (1881)

- *Sporotrichum sulfurescens* Van Beyma (1928)
- *Stachylidium bassianum* (Bals.-Criv.) Mont. (1856)
- *Trichoderma minimum* (Speg.) G. Müller (1965)
- *Tritirachium shiotae* (Kuru) Langeron (1947)

Local Common Names

- **Germany:** Parasit: Arthropoden

EPPO code

- **BEAUBA (*Beauveria bassiana*)**

Taxonomic Tree

- Domain: Eukaryota
- Kingdom: Fungi
- Phylum: Ascomycota
- Subphylum: Pezizomycotina
- Class: Sordariomycetes
- Subclass: Hypocreomycetidae
- Order: Hypocreales
- Family: Cordycipitaceae
- Genus: Beauveria
- Species: *Beauveria bassiana*

- Exploration and identification

The specific term is given in honor of Agostino Bassi, an entomologist from Italy who identified it in 1835 as the underlying factor of the muscardine illness in domesticated silkworms. At one point in the past, it was also referred to as Tritirachium shiotae.

The structure of the fungi

B. bassiana develops as a white mold in culture. It generates numerous dry and powdery conidia in unique white spore balls on typical cultural media. These spore balls consist of a cluster of short and ovoid conidiogenous cells, ending in a narrow apical extension called a rachis. The rachis elongates after each conidium formation, producing a lengthy zig-zag extension. The conidia are individual, haploid, and water-repellent.

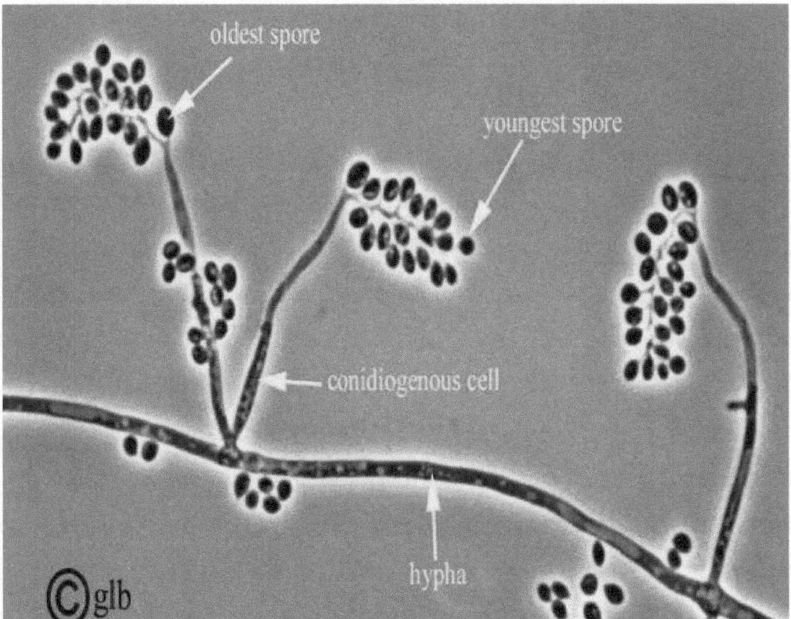

Fig. (6): ***Beauveria bassiana***

Classification of Entomopathogenic Fungi

Class: plant kingdom

Mastigomycotina

Class: Deuteromyceteae

Or : Moniliales

Fam : Moiliaceae

Beauveria & Metarhizium

White muscardine disease

White muscardine disease is a condition characterized by its presence in white color.

White muscardine disease is the term used to describe the insect disease caused by the fungus. When the small spores of the fungus come into contact with an insect's body, they germinate, penetrate the outer layer, and grow internally, resulting in the death of the insect within a few days. Subsequently, a white mold appears on the dead body and produces new spores. Different isolates of B. bassiana can target a wide variety of insects, and their ability to do so may vary. The factors responsible for making certain insects more susceptible to the disease are not currently understood. Researchers have observed that Beauveria bassiana, when parasitizing the Colorado potato beetle, can become a host itself to a mycoparasitic fungus called Syspastospora parasitica. This particular fungus also a ttacks insect-pathogenic species from the Clavicipitaceae family.

Fig. (7): ***Beauveria bassiana*****, Mass rearing on Rice medium** [131]

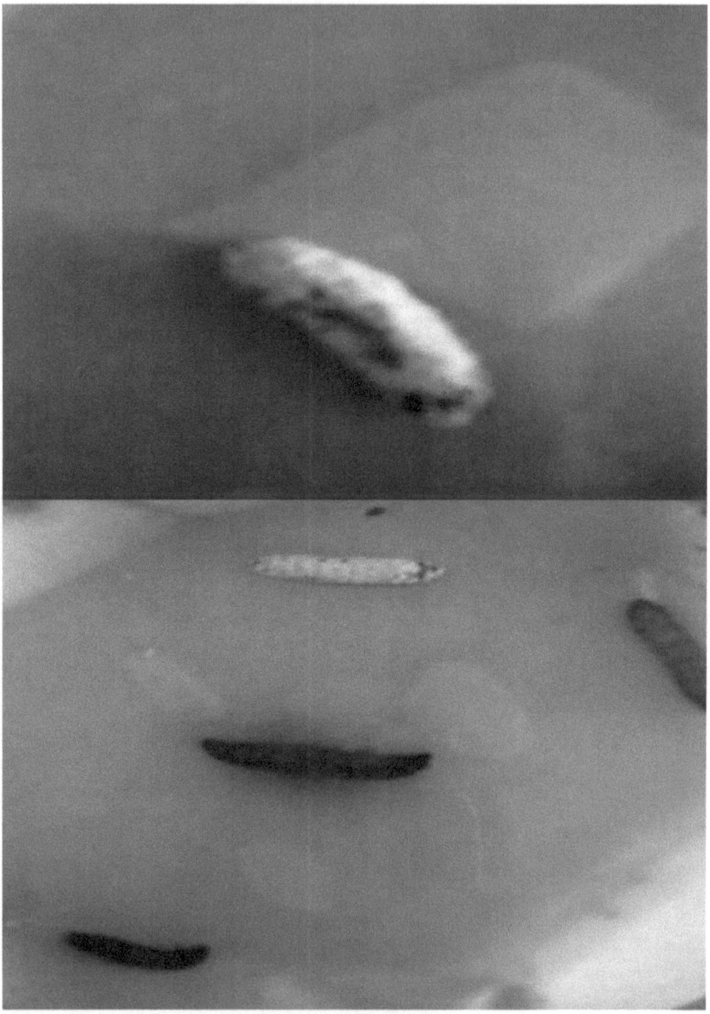

Fig. (8): ***Beauveria bassiana*****, infection insects,** [132, 133]

The life cycle of *B. bassiana* involves two distinct modes of growth

In the absence of its specific insect host, Beauveria goes through an asexual vegetative life cycle, which includes germination, filamentous growth, and the creation of sympodulo conidia. However, when the host insect is present, Beauveria switches to its pathogenic life cycle. The conidiospores germinate on the surface of the insect's cuticle, and the germinated hyphal tubes penetrate the integument directly. Once inside, the fungus undergoes a change in growth morphology to a yeastlike phase and produces hyphal bodies that circulate in the insect's haemolymph and multiply through budding. After the host insect's death, the fungal growth returns to its typical hyphal form, known as the saprotrophic stage. The ability to transition to the yeast-
like phase might be essential for the fungus to cause disease.

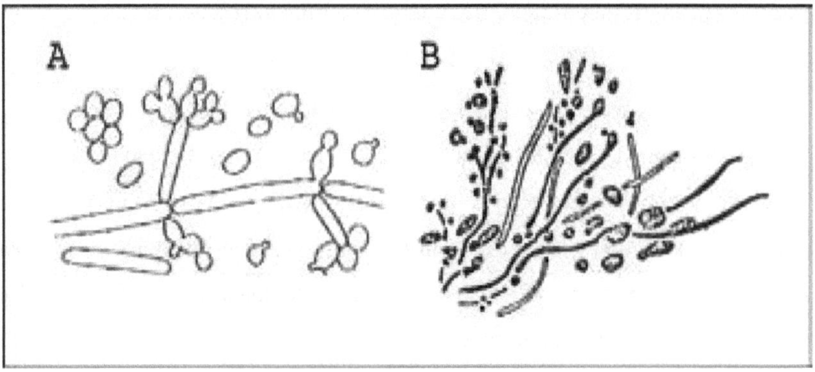

Fig. (9):

The growth pattern of B. bassiana can be classified into two distinct forms. During the yeast phase, it behaves similarly to a parasite when it infects susceptible species. On the other hand, during the saprobic phase, it exhibits filamentous hyphae.

Counting the number of spores

The spore suspension was completely mixed using a vortex and then diluted with sterile 0.05% (v/v) Tween 80.

The number of spores in the diluted solution was determined by using a Hemacytometer (Superior, Marienfeeld, Germany) and a light microscope (Olympus BH2, Japan) at a magnification of × 400.

The process of infection

Unlike bacteria and viruses, which enter the gut wall from contaminated food, fungi have a distinct method of infection. They make their way into the hamocoel by penetrating the cuticle or possibly entering through the insect's mouth parts. When fungal spores are swallowed, they do not sprout in the digestive system and are excreted in the feces.

The insect's death is caused by a combination of factors: physical harm from tissue invasion, depletion of nutrients, and poisoning.

Attachment and sprouting of fungal spores

The initial step in the establishment of mycosis is the attachment of a fungal spore to the surface of a susceptible host. For entomopathogenic fungi, this attachment is typically a random process facilitated by wind or water.

It has been observed that dry spores of B. bassiana have an outer layer composed of intertwined hydrophobic rodlets. This rodlet layer is unique to the conidial stage and is not present on vegetative cells.

The adhesion of dry spores to the cuticle is believed to be caused by the non-

specific hydrophobic forces exerted by the rodlets [31].
Additionally, lectins, a type of carbohydrate-binding glycoproteins, have been found on the surface of B. bassiana conidia.
It has been suggested that lectins may play a role in the binding between conidia and the insect cuticle. However, the exact mechanisms of the interaction between fungal spores and the cuticle remain unknown [32].
Once the pathogen reaches and attaches to the host surface, it undergoes rapid germination and growth. These processes are greatly influenced by factors such as nutrient availability, oxygen, water, pH, temperature, and the presence of toxic compounds on the host surface.
Fungi with a wide range of hosts typically germinate in culture in response to a variety of nonspecific carbon and nitrogen sources.
On the other hand, entomopathogenic fungi with a limited host range seem to have more specific requirements for germination [33].

Development of a structure for infection

Entomopathogenic fungi enter their hosts by directly penetrating the cuticle of the host. The cuticle consists of two layers: the outer epicuticle and the procuticle.

The epicuticle is a complex and thin structure that does not contain chitin but instead contains proteins that are stabilized by phenol. It is also covered by a waxy layer that contains fatty acids, lipids, and sterols.

The procuticle makes up most of the cuticle and contains chitin fibrils that are embedded in a protein matrix along with lipids and quinones. Protein can make up as much as 70% of the cuticle. In certain areas of the cuticle, the chitin is arranged in a helical manner, creating a laminate structure.

Like many other entomopathogenic fungi, *B. bassiana* conidia germinate on the surface of the host and develop an infection structure called an appressorium.

The penetration of the outer layer of the skin

Pathogenic fungi must penetrate the insect body through the cuticle in order to obtain nutrients for their growth and reproduction. The process of entering the host involves both enzymic degradation and mechanical pressure, as evidenced by the physical separation of lamellae caused by the hyphae. There are various extracellular enzymes, such as chitinases, lipases, esterases, and different classes of proteases, that are believed to play a role in the pathogenesis of fungi by degrading the major components of the insect cuticle. The production of cuticle-degrading enzymes by M. anisopliae during infection structure formation has been studied through biochemical and histochemical analysis, both in vivo and in vitro, using Calliphor vomitoria and Manduca sexta. Endoproteases (PR1 and PR2) and a minopeptidases are among the first enzymes produced on the cuticle during the for mation of appressoria. N-
Acetylglucosaminidase is produced at a slower rate compared to the proteolytic enzymes. Chitinase and lipase activities were not detected. Despite the complex structure of the insect cuticle, synergy is likely required for penetration.

Toxin generation

Fungal toxins and their role in host death have been supported by substantial indirect evidence provided by deuteromycete pathogens. The infiltration of hyphae is preceded by the disruption of cells, indicating the potential action of cytotoxins. Various behavioral symptoms observed in mycosed insects, including partial or complete paralysis, decreased activity, and reduced irritability, align with the effect s expected from neuromuscular toxins [42].

B. bassiana and M. anisopliae have been found to produce significant quantities of toxic substances within their hosts.

As examples, Beauvericin, Beauverolides, Bassianolide, and Isarolides are some of the toxins identified.

Can refer to the systems within a host organism that are responsible for defending against harmful pathogens or foreign substances.

To prevent invasion by fungi, insects have developed different defense mechanism s. These mechanisms consist of both passive structural barriers, such as the cuticle, and a series of active responses to pathogens that enter the hemocoel.

Active responses include melanization, cellular reactions, humoral reaction to ident ify foreign pathogens, and the production of protease inhibitors.

Melanization, which involves the oxidation of phenolic compounds to dihydroxyph enylalanine and the production of brown or black melanic pigments, is a common way for insects to respond to fungal infections.

Melanin can partially protect the cuticle from enzymatic attack or be toxic to fungi. However, this protection is not complete. Research from [32] suggests that melani

zation is mainly effective against weaker or slower-
growing pathogens, but is ineffective against more virulent fungi.

Cellular reactions occur once the cuticle and epidermis are breached by the invading fungus. The defense systems of the hemolymph are then encountered by t he fungus.
Mycopathogens within the haemocoel trigger responses such as phagocytosis, encapsulation, and nodulation. However, the impact on fungal elements remains uncertain. Through the arbitrary injection method, it was discovered that haemocytes of the migratory grasshopper, *M. sanguinipes*, encapsulate viable conidia of *B. bassiana*, but are unable to prevent conidial germination within the nodule.
It was hypothesized that the release of toxins and extracellular proteases by *B. bassiana* might induce the evasion of encapsulation.

Manufacture of protease inhibitors:
Hostderived protease inhibitors may play a role in insect defense mechanisms by hindering the activities of pathogens that degrade the insect's outer protective layer . These inhibitors have been extracted from the blood of Anticarsia gemmatalis larvae, which showed resistance to infection caused by Nomuraea rileyi [38].
Various aspects of employing entomopathogenic fungi as biological control agents.

(1) One of the advantages of utilizing fungi as insecticides is their ability to specifically target and control harmful insect pests while leaving beneficial insect predators and nonharmful parasites unharmed.

(2) Fungi do not have any detrimental effects on mammals, thereby reducing the typical hazards associated with the use of insecticides, such as environmental pollution.
(3) Insect resistance and prolonged pest control are not issues when using fungi as insecticides.
(4) There is a great potential for further advancement through biotechnological research in the field of using fungi as insecticides.
(5) Entomopathogenic fungi have a high persistence in the environment, which allo ws for long-term effects on the suppression of pests.

However, there are also several limitations to using fungi as insecticides:
(1) It takes approximately 2-
3 weeks for fungi to kill insects, whereas chemical insecticides might only require 2-3 hours.
(2) The application of fungi needs to coincide with periods of high relative humidity, low pest numbers, and an absence of fungicides.

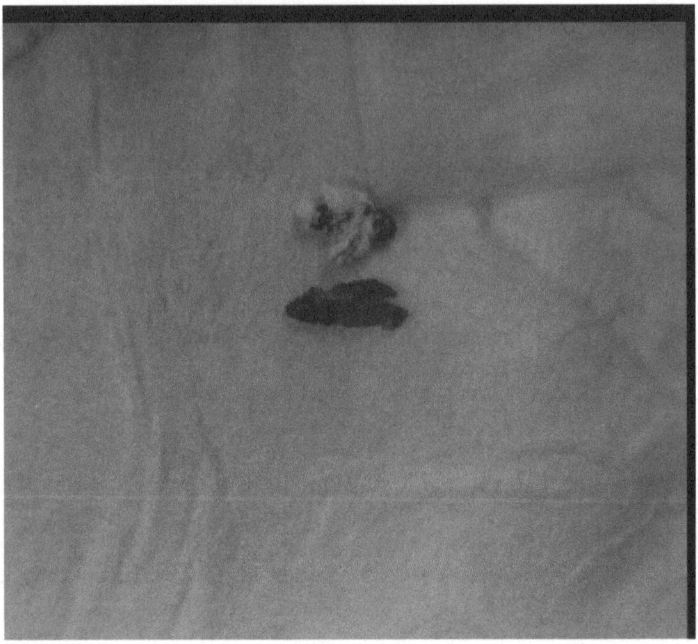

Fig. (10): Pupa of *S. ocellatella* infected with *B. bassiana* [12]

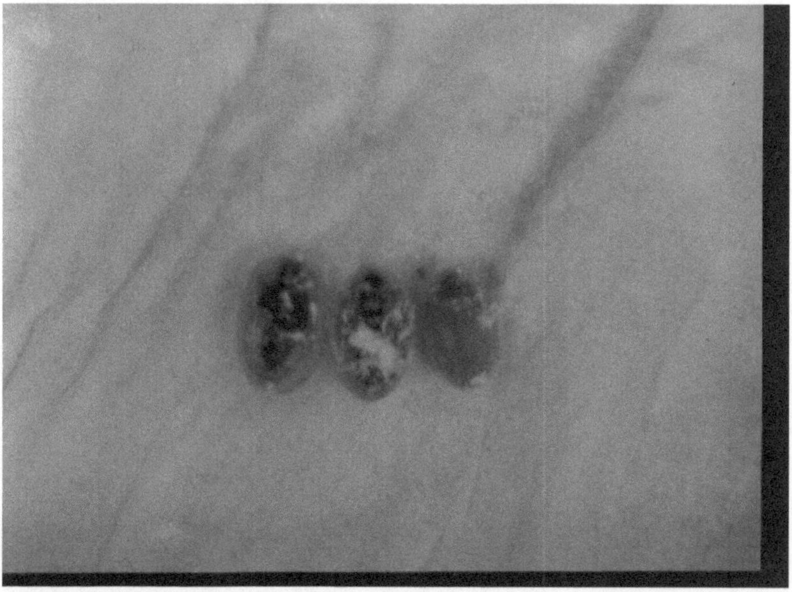

Fig. (11): Adult of *Cassida vittata* infected with. *B. bassiana* [12]

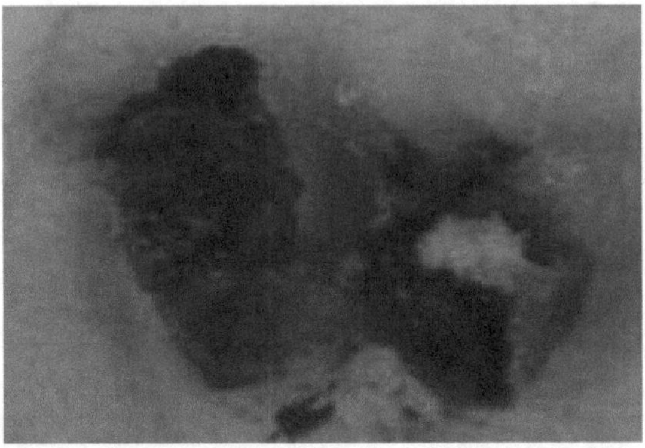

Fig. (12): Pupa of *C. vittata* infected with *B. bassiana* [12]

Fig. (13):

The natural enemy of the red palm weevil, Rhynchophorus ferrugineus, is shown i n both dorsal and ventral views. These images display Beauveria bassiana, which i s seen sporulating on the dead body of the weevil.

Manipulating the genes of fungi that are pathogenic to insects through genetic engineering

The broader use of fungi for biological control will rely on enhancing the character istics of wildtype strains through the combination of different strains and mutants. There are two types of enhancements that can be considered: (i) improving the effectiveness of the insecticide, such as reducing the required dosage to kill insects, reducing the time required to eliminate the pests, or decreasing crop damage cause d by the pests by reducing feeding time; (ii) expanding the range of host organisms affected.

A comprehensive understanding of the remarkable pathology of fungal infections is crucial for the development of a highly virulent strain. Molecular biology provides the necessary tools for investigating the mechanisms of pathogenesis and, in the long term, for creating recombinant organisms with new characteristics. Initial progress toward these objectives has been made with M. anisophliae, and to a lesser extent with B. bassiana [39].

Genetic transformation systems, which are integral to modern fungal research and essential for the experimental manipulation of virulence genes in both controlled laboratory settings and real-world environments, have been successfully established [40].

The success of these techniques relies on having selectable transformation markers available.

Transformation methods have been employed to isolate specific pathogenic genes, explore the determinants of virulence in M. anisopliae, and generate a strain with enhanced virulence.

Decoding the molecular mechanisms of fungal pathogenesis in insects will serve as the foundation for genetically engineering entomopathogenic fungi.

Insect biological control application

Beauveria bassiana, which has the ability to control various pests including termites, whiteflies, and other insects, can be utilized as a biological insecticide. It s potential in managing mosquitos that transmit malaria is being studied.

When used as an insecticide, the spores of Beauveria bassiana are either sprayed on affected crops in the form of an emulsified suspension or wettable powder, or applied onto mosquito nets as a means of mosquito control.

Due to its broad range of potential arthropod hosts, Beauveria bassiana should be considered a nonselective biological insecticide. Hence, it should not be employed on flowers that are frequented by pollinating insects.

Identified objectives consist of

Various types of insects including aphids, whiteflies, mealybugs, psyllids, Lygus bugs, chinch bugs, grasshoppers, thrips, termites, fire ants, flies, stem borers, fungal gnats, shoreflies, beetles, coffee borer beetles, Colorado potato beetles, Mexican bean beetles, Japanese beetles, Boll weevils, cereal leaf beetles, bark beetles, black vine weevils, strawberry root weevils, caterpillars, European corn borers, codling moths, Douglas fir tussock moths, silkworms, and mites pose potential threats to crops and plants.

There have been rare instances of infection by *B. bassiana*, a fungus used as an insecticide, in humans with suppressed immune systems, although it generally does not infect humans or other animals. The use of *B. bassiana* as an insecticide is considered safe. However, caution should be taken as the spores can worsen respiratory conditions.

Wagner and Lewis conducted a study on the ability of B. bassiana to grow as an endophyte in corn.

It is challenging to create a detailed slide mount showing the fine details of the origin and development (ontogeny) of the spores in Beauveria due to the small and delicate nature of its spore producing apparatus.

The photomicrograph above illustrates three conidiogenous cells (spore mother cells) on a hypha (fungal filament) of Beauveria bassiana. The asexual spores (con idia) are produced in a youngest-to-
oldest formation on an elongating conidiogenous cell (cell that produces conidia).

Conidium is the term used to describe a non-
sexual spore that germinates through a germ tube, which is produced in the Hypho mycetes class of the Deuteromycota division for asexual fungi.

Beauveria's method of generating spores through sympodial growth

The spores in *B. bassiana* are produced in a sympodial manner. A spore is generate d at the end of the mother cell, which leads to the cessation of growth in the mother cell. A new growing point emerges just below this terminal spore, grows past it, and then produces a second spore at a higher level. With the consumption of the new growing point, a third growing point is subsequently initiated just below the second spore. With each spore production, the hyphal tip is depleted and a new growing point is generated. This process results in a sequence of spores, with the youngest spore located at the tip, known as the acropetal succession, and the spore head becoming progressively longer. Once all the spores are dislodged, the spore-bearing tip of the conidiogenous cell assumes a zigzag appearance and is referred to as a rachis (this term is also used to describe the seed head in wheat when all the seeds are gone).

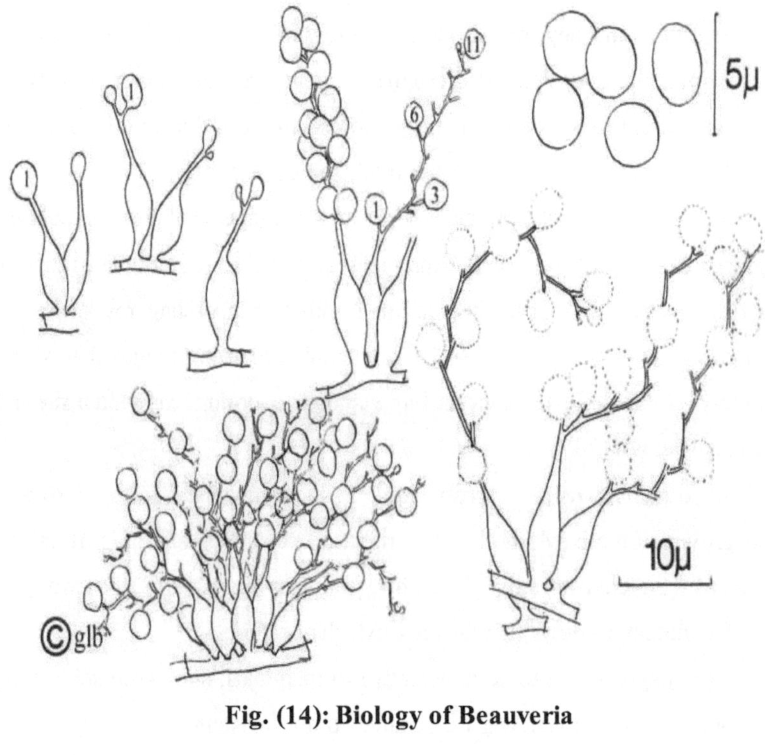

Fig. (14): Biology of Beauveria

Beauveria bassiana is an aggressive parasite of many different insect host species. Not only does it have a wide host range but insects are attacked at larval or adult stages.

The spores are tiny, measuring only a few microns. The hyphae (hyaline) and so and spores are non pigmented colonies appear white in cultures or tufts of white mycelium bearing masses of powdery spores burst out through the body parts of infected insects as in the cicada, hawk moth and chinch bugs.

For something really dramatic look at the synnematal form of *Beauveria* I found growing on a hawk moth. And right next to that is a couple of hairy chinch bugs that I recovered from the chinch bug population inhabiting my lawn.

Beauveria bassiana strain 447 is a naturally occurring fungus that is approved for indoor use as an insecticide in bait stations to control individual fire ants and other ants (but not colonies.)

For something really dramatic look at the synnematal form of *Beauveria* I found growing on a hawk moth. And right next to that is a couple of hairy chinch bugs that I recovered from the chinch bug population inhabiting my lawn.

- The fungal spores infect the ants ‹which soon die.
- The fungus is not toxic or infective to mammals, and exposure to the public and the environment will be minimal to non-existent.
- Therefore, no adverse effects are expected on children, adults, pets, or the environment when the bait stations are used as directed.

Description of the Active Ingredient

- *Beauveria bassiana* strain 447 belongs to a group of fungi that are found throughout the world.
- The infection begins when a fungal spore starts growing on the outer skeleton of a host insect, and the infected insect soon dies.

Assessing Risks to Human Health

- Whether or not a substance poses a risk to humans or other organisms depends on two factors:
- How toxic the substance is, and how much of it an organism is exposed to.
- Therefore, the EPA considers toxicity data and exposure data in determining whether to approve a pesticide for use.

No adverse human health effects are expected from use of *Beauveria bassiana* strain 447 in bait stations.

- The bait stations are designed to prevent adults, children, or pets from being exposed to the fungus.
- In addition, laboratory studies show that the active ingredient is not toxic or infective to laboratory animals when eaten or inhaled.
- The fungus grows only at temperatures lower than human body temperature, so no infection is expected.

Assessing Risks to the Environment

- No harmful environmental effects are expected because the bait stations are approved only for use indoors.
- Because the active ingredient is firmly attached to the inside of the bait station, no outdoor exposure is expected.
- EPA will require additional studies if the bait stations are proposed for outdoor or other additional uses.

Products Directed Against Public Health Pests

- EPA defines a public health pest as any organism that can cause or transmit human disease, or can cause human discomfort or injury.
- Fire ants are categorized as a public health pest.
- To help protect the public's health, EPA requires registrants of products used against public health pests to demonstrate that the product meets specific standards for effectiveness as well as for safety.

At present improved fungal mass production and formulations are commercially available.

Modern techniques in genetic engineering and biotechnology are extremely helpful in manipulating the desired traits of entomopathogenic fungi, which can further improve its bioactivity·

Entomopathogenic fungi were among the first organisms to be used for the biological control of pests. More than 700 species of fungi from around 90 genera are pathogenic to insects.

Most are found within the deuteromycetes and entomophthorales. some insect-pathogenic fungi have restricted host ranges, for example, *Aschersonia aleyrodis* infects only scale insects and whiteflies, while other fungal species have a wide host range, with individual isolates being more specific, for example, *Metarhizium anisopliae* and *Beauveria bassiana*.

The commercial produced entomopathogenic fungi and their targeted hosts are shown in Table 4.

Fungal species such as *M. anisopliae* and *B. bassiana* are well characterized in respect to pathogenicity to several insects and they have been used as agents for the biological control of agriculture pests worldwide.

In Colombia, about 11 companies offer at least 16 products based on the entomopathogenic fungi *B. bassiana*. These products are used not only in the coffee crop but also in other crops such as cabbage, corn, bean, tomato, potato. They are also used to treat public disease vectors (e.g., flies and mosquitoes) [41].

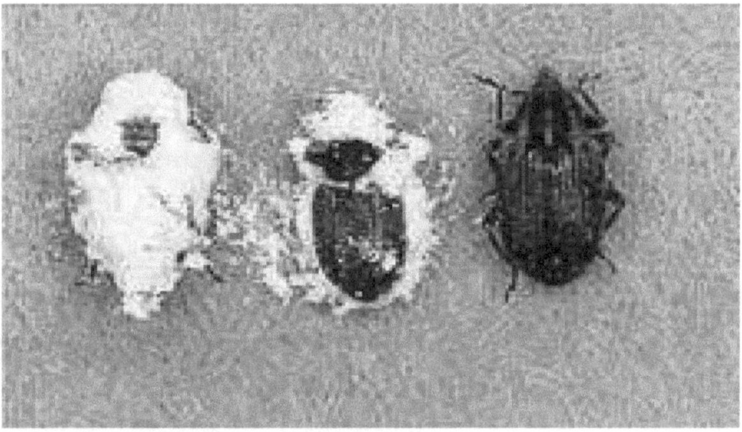

Fig. (15): FiBiological control of insect pest (alfalfa weevil). Left and Center, Adult insect killed by *Beauveria bassiana*. Right, Healthy adult insect (control).

Table (4): Entomopathogenic fungi in commercial and experimental production (adapted from Khachatourians, 1986)

Fungus	Targets	Product
Beauveria bassiana	Colorado potato beetle Coding moth European corn borer Pine caterpillar	Boverin Mycotrol®
Culicinomyces clavisporus	Mosquito larvae	
Hirsutella thompsonii	Citrus rust mite	Mycar
Metarhizium anisopliae	Spittle bug Sugarcane frog hopper	Metaquino Meta-Sin®
Verticillium lecanii	Aphids Coffee green bug Greenhouse whitefly Thrips	Vertalec

Metarhizium species

- *Metarhizium* species is being tested as natural enemy of many soil insect pests.
- *Metarhizium* can be an important control agent of aphids.
- In Indonesia (West Sumatra), *Metarhizium* is used as a spraying solution for control of different pest insects in chili.
- In Philippines, *Metarhizium* effectively reduced populations of rhinoceros beetles in coconut [45].
- In the United States *Metarhizium* is commercially available as a foliar spray.

Metarhizium is usually called the green fungus of insects. The fungus has white mycelia within the body, but when it is ready to form spores, the spores coat the killed host with a velvety covering of olive-green.

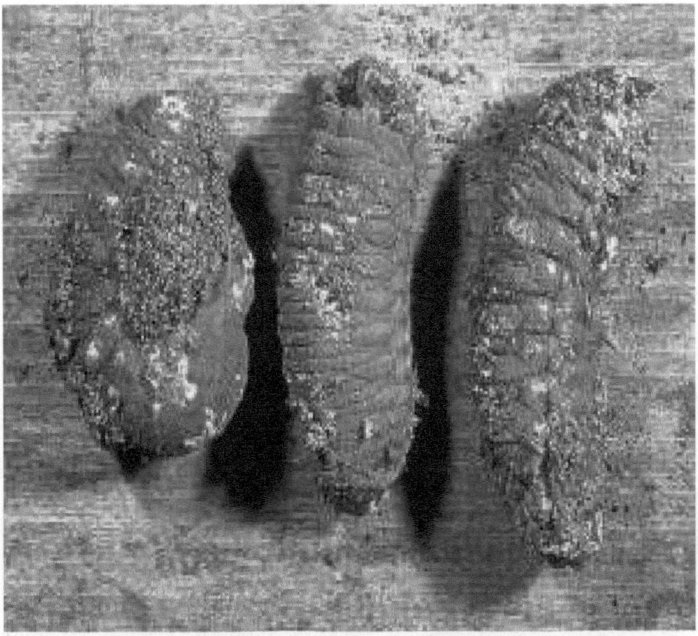

Fig.(16): Green June beetle grub skilled by *Metarhizium*.

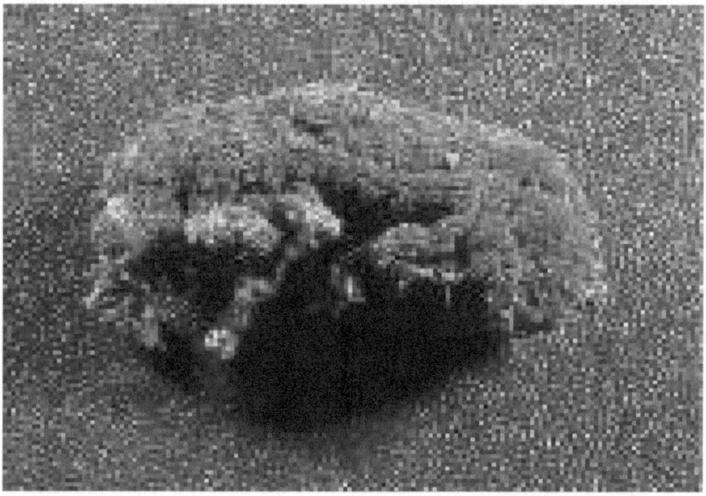

Fig.(17): Japanese beetle grub infected with *Metarhizium*.

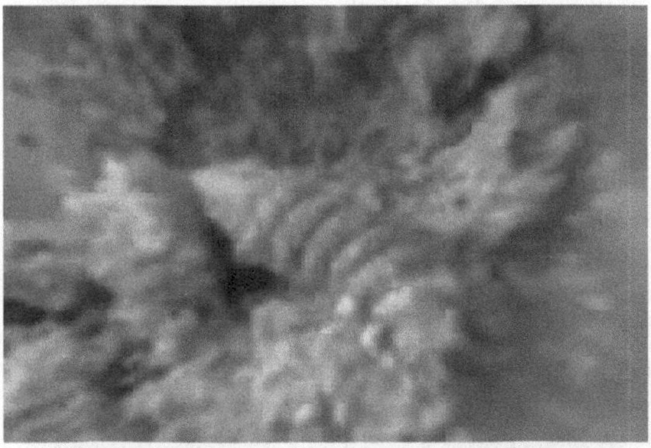

Fig.(18): ***Agrotis ipsilon*** **with** ***Metarhizium*** **[46]**

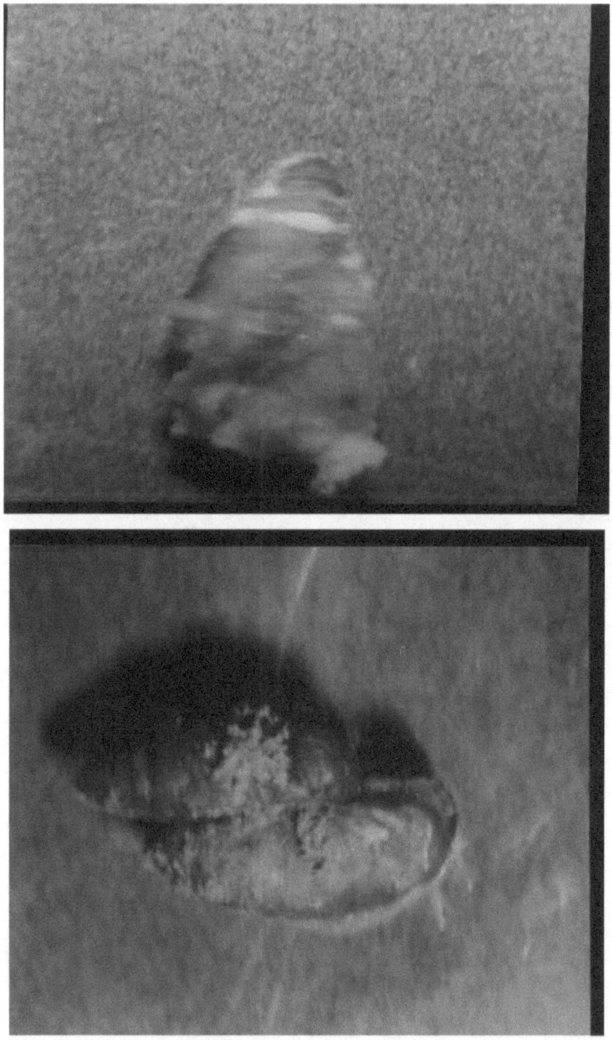

Fig.(19): Pupa of *S. ocellatella* infected with *Metarhizium* [47]

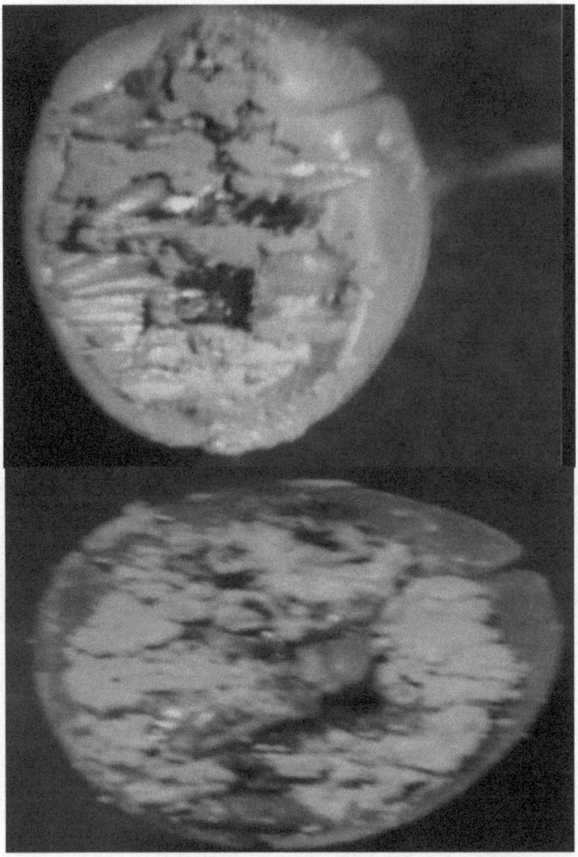

Fig.(20): Adult of *C. vittata* infected with *Metarhizium* [47]

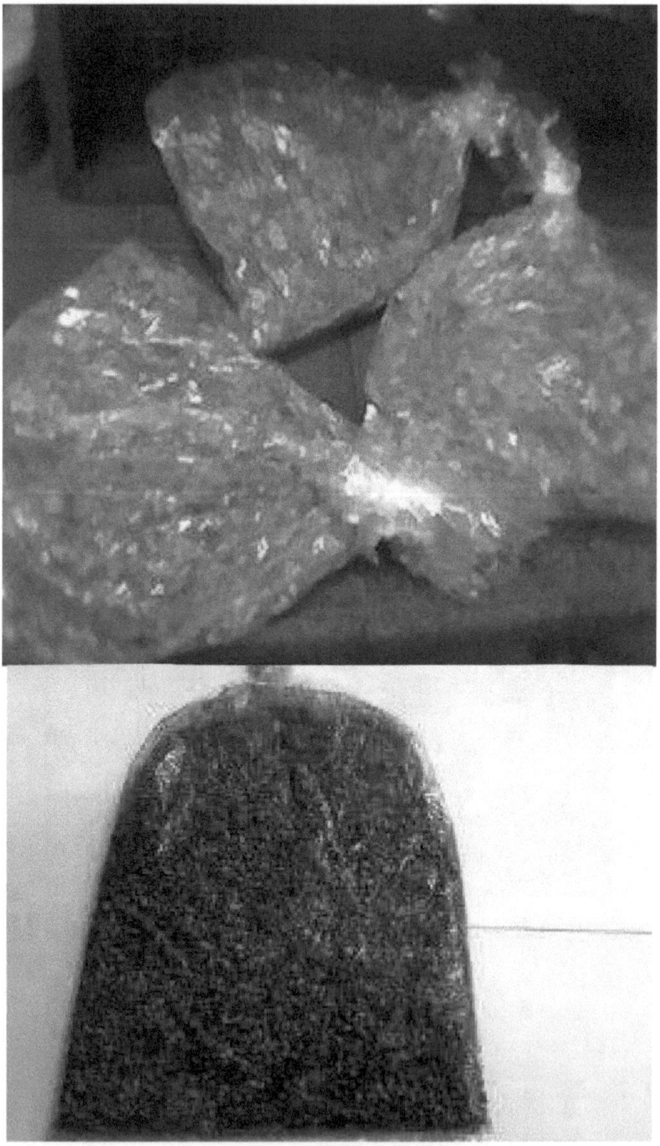

Fig.(21): Rice grains infected with *Metarhizium* [48]

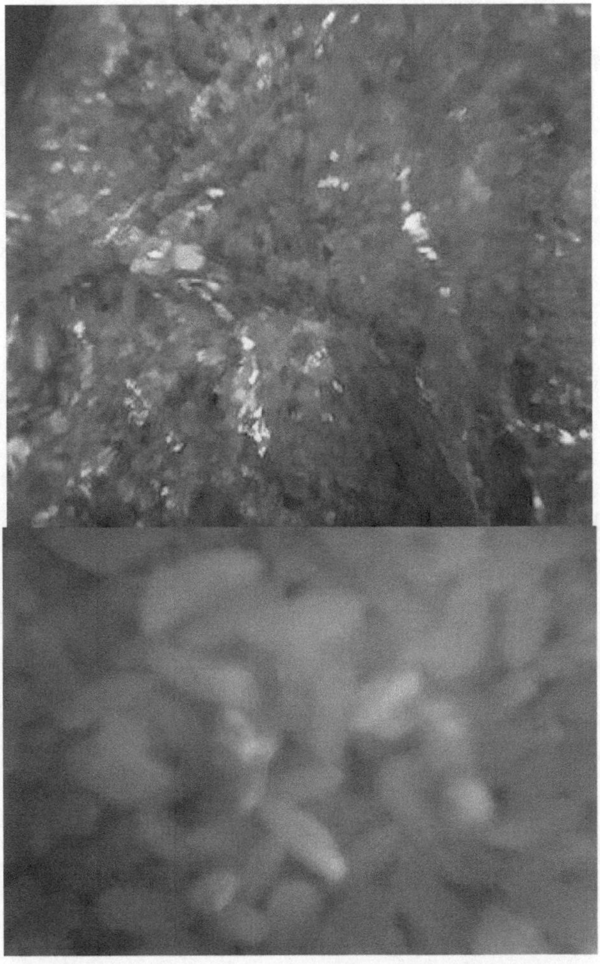

Fig.(22): Rice grains infected with *Metarhizium* [48]

Mode of action and symptoms

- Fungi penetrate the skin of insects.
- Once inside the insect, the fungus rapidly grows throughout the body.
- Death is caused by injury to the tissue or, occasionally, by toxins produced by the fungus.
- The fungus grows out through the insect's skin to produce spores that can sometimes be seen as "powder" or "cotton" on the skin of the dead insect.

- When spread by wind, rain or contact with other insects, the spores can spread the infection.
- Infected insects stop feeding and become lethargic.
- They may rapidly die, sometimes in a position still attached to a leaf or stem.
- The dead insect's body may be firm or it may be an empty shell.
- The fungus is often seen as "hairs" or "cotton" in various colors on parts of the insects body.

Effectiveness

- Insect-pathogenic fungi usually need high humidity to cause infection.
- Natural infections are therefore most common during the wet season.
- The effectiveness of fungi against insect pests depends on many factors:

Having the correct fungal species with the susceptible insect life stage, at the appropriate humidity, soil texture (if you are trying to reach ground-dwelling insects), and temperature.

The fungal spores, which can be carried by wind or water, must contact the pest insect to cause infection.

- When insect-pathogenic fungi are applied, for example through a spore-solution, good spray coverage of the plants is essential.

How to protect and help fungi

- Many insect-pathogenic fungi live in the soil.
- There is evidence that application of some soil insecticides, fungicides and herbicides can inhibit or kill these fungi.
- For example, even very low concentrations of some herbicides can severely limit the germination and growth of *Beauveria bassiana* fungal spores in soil samples.
-

Primary or secondary

- When you leave a plate of food with for example some chicken meat in your kitchen and leave it for a few days, you may find the meat covered with fungus when you look again.
- This fungus is called a *secondary infection*:
- It was not the reason the chicken died (the chicken probably died because a person killed it), and the fungus started growing after the chicken was dead.

Similarly, when insects are dead, some fungi may start growing on the dead insect.

- This is also called a secondary infection.
- These fungi that cause secondary infection are part of the "trash men team" of nature:
- They make sure dead things are decomposed quickly.
- When the fungi actually cause a living insect to die, like insect-pathogenic fungi do,
- It is called a *primary infection*

Interesting, but who cares?

- When searching for beneficial fungi in the field, it is important to distinguish between fungi causing primary infection (pathogens) and those causing a secondary infection.
- Fungi that cause secondary infections are not really pathogens (will not infect and Kill healthy insects).

Application of Entomopathogenic Fungi for Insect Pests Control

Bioassay of isolated fungi against Many Insects:

Larvae of, *G. mellonella*

G. mellonella is usually used as a test insect for assessment of virulence of entomopathogenic microorganisms including fungi, [49]. Fungal concentrations around $1x10^7$ spores/ml are frequently used in these bioassayes. Target insect may be directly used if available. Gupta, [49], compared the virulence of five strains of *B. bassiana* against the larvae of *G. mellonella*.

% of mortality among *G. mellonella* larvae exposed for 10 days to serial concentrations were ((C_1) $1x10^3$, (C_2) $1x10^4$ and (C_3) $1x10^5$ spores /ml.). *B. bassiana* the most virulent fungus to *G. mellonella* larvae followed by *M. anisopliae* then *V. lecanii*. At the lowest concentration of (1x 10^3 spores/ ml) mortality of *G. mellonella* larvae ranged 14.4 to 99.2 %, 10.4 to 90.4% and 10.4 to 84% due to *B. bassiana, M. anisopliae* and *V. lecanii,* respectively. At the highest concentration, the three fungi reached their maximum effect at 10 days after application being 100%, 98.4% and 97.6%, respectively, [26]. Virulence depended greatly on the concentration and the exposure time.

B. bassiana the fastest effect against larvae of *G.mellonella*, followed by *M. anisopliae* then *V. lacanii*. Strong relationship was found between either the concentration of the fungus or the time of exposure and the resulted mortality in larvae of *G. mellonella*. Values of squared correlation coefficient (R^2) were always over 0.9, [26].

Using of *B. bassiana* and *M. anisopliae* were the most effect to decrease the total number of *C. vittata* (Vill.) The fungus *B. bassiana* was the most effect to controlling the tortoise beetle, *C. vittata* than the Fungus *M. anisopliae.*

Table (5) the susceptibility of different stages of tortoise beetle *C. vittata* to entomopathogenic fungus *B. bassiana at* different concentrations, [12]:

Treated stages	%Mortality at indicated concentrations					LC 50	Fudicial limits 95 %	Slope ± SE
	10^6	$5x10^6$	10^7	$5x10^7$	10^8			
1st instar larvae	50	75	85	95	100	1.08×10^6	($6.35x10^5$ -1.6×10^6)	1.09± 0.121.
2nd instar larvae	40	55	80	90	100	2.24×10^6	($2.92x10^5$ –$5.42x10^6$)	1.1 ± 0.2
3rd instar larvae	40	50	65	90	100	$1.49x 10^6$		0.79± 0.31
4th instar larvae	80	80	85	80	100	$1.03x10^5$	($0.2x10^5$ –$602x10^5$)	2.33 ±0.73
5th instar larvae	50	60	70	70	85	$1.04x 10^6$	($2.16 x10^5$ –$2.38x10^6$)	0.43 ± 0.08
Pupa	50	65	65	75	95	$1.32x10^6$	($0.013x10^6$ –$6.04x10^6$)	0.59 ±0.17
Adult	80	85	85	95	100	$0.4x10^5$	($0.005x10^5$ –$3.13 x10^5$)	2.55 ± 0.76

Table (6) the susceptibility of different stages of tortoise beetle *C.vittata* to entomopathogenic fungi *M. anisopliae at* different concentrations, [12]:

Treated Stages	%Mortality at indicated concentration					LC $_{50}$	Fudicial limits 95 %	Slope ±SE
	10^6	5x 10^6	10^7	5x10^7	10^8			
1st instar larva	60	70	70	95	100	3.68 x10^6		1.4± 0.5
2nd instar larva	60	55	55	60	75	2.45x10^5	(6.86x10^5- 1.2x 10^6)	0.16 ± 0.08
3rd instar larva	60	65	70	85	100	8.49 x10^5		0.69 ± 0.22
4th instar larva	60	60	60	70	100	9.8 x10^5		0.52 ± 0.49
5th instar larva	55	55	65	75	90	1.22 x10^6	(0.03x10^6 - 5.79x10^6)	0.5 ± 0.14
Pupa	45	50	55	60	70	3.65 x10^6	(6.32 x10^5 – 8.72x10^6)	0.3 ± 0.08
Adult	60	75	70	95	100	2.27 x10^5	(0.9x10^5 – 1.12 x10^6)	2.15 ± 0.76

Table (7) the susceptibility of different stages of sugar beet mining moth *S. ocellatella* to entomopathogenic fungus *B. bassiana at* different concentrations, [12]:

Treated stages	%Mortality at indicated concentrations					LC_{50}	Fudicial limits 95 %	Slope ± SE
	10^6	$5x10^6$	10^7	$5x10^7$	10^8			
4th instar larva	60	75	60	60	70	$0.4 x10^5$		0.04 ± 0.14
Pupa	75	90	95	100	100	$2.99 x10^5$	($0.9 x10^5$ –5.63 x 10^5)	1.2 ± 0.19

Table (8) the susceptibility of different stages of sugar beet mining moth *S. ocellatella* to entomopathogenic fungi *M. anisopliae at* different concentrations, [12]

Treated stages	%Mortality at indicated concentrations					LC_{50}	Fudicial limits 95 %	Slope ± SE
	10^6	$5x10^6$	10^7	$5x10^7$	10^8			
4th instar larva	70	70	75	90	95	$2.27x10^5$	($6.14 x10^5$ – $6.53 x10^5$)	0.52 ± 0.09
Pupa	20	20	35	70	75	$1.92 x10^7$	(6.14 x10 6 –$9.6 x10^7$)	0.88 ± 0.18

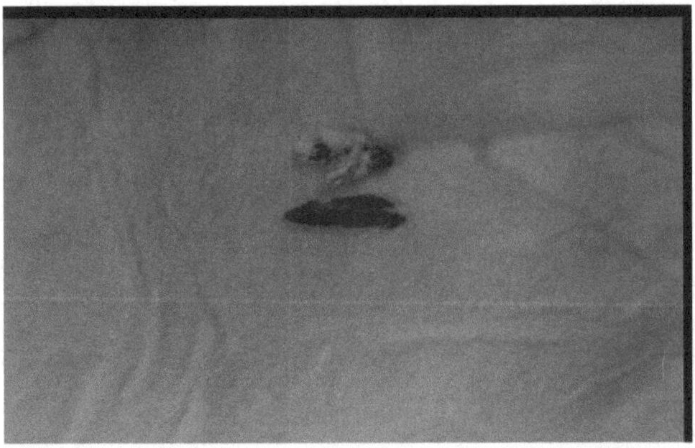

Fig . (23): Pupa of *S. ocellatella* infected with *B. bassiana*, [12]

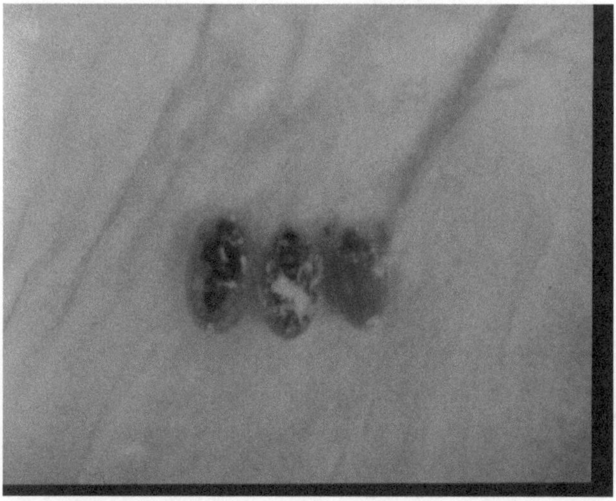

Fig. (24): Adult of *C. vittata* infected with. *B. bassiana*, [12]

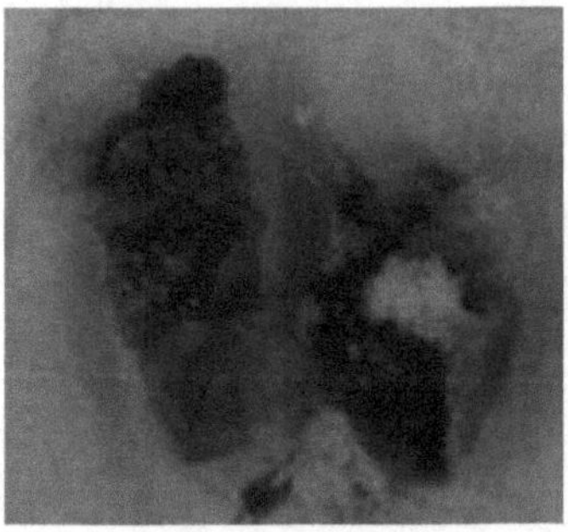

Fig. (25): Pupa of *C. vittata* infected with *B. bassiana*, [12]

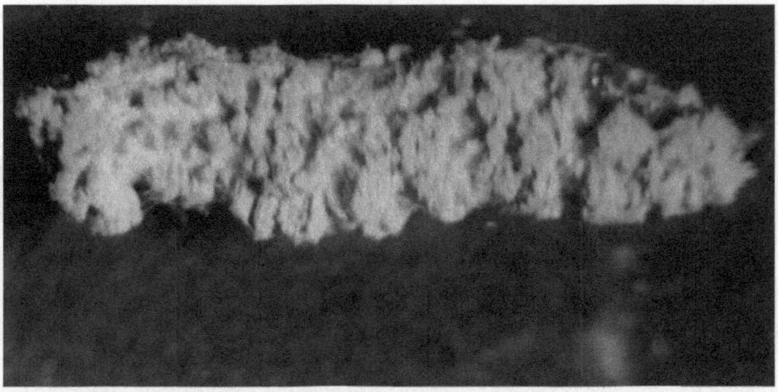

Fig. (26): Larvae of *S. ocellatella* infected with *M. anisopliae*, [12]

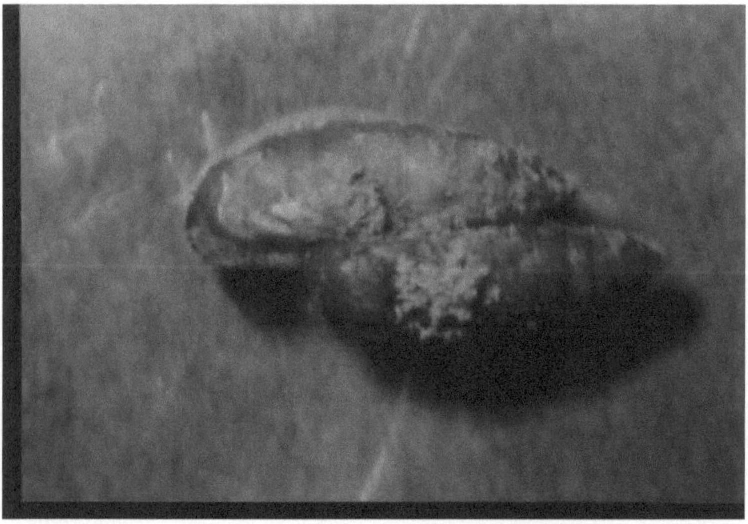

Fig . (27): Pupa of *S. ocellatella* infected with *M. anisopliae*, [12]

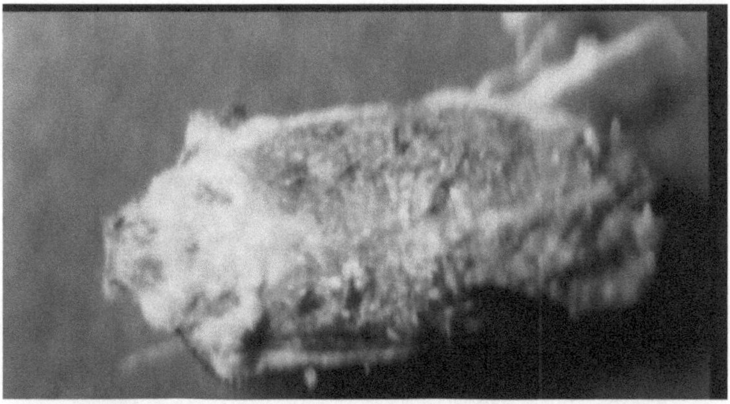

Fig. (28): Adult of *S. ocellatella* infected with M. *anisopliae*, [12]

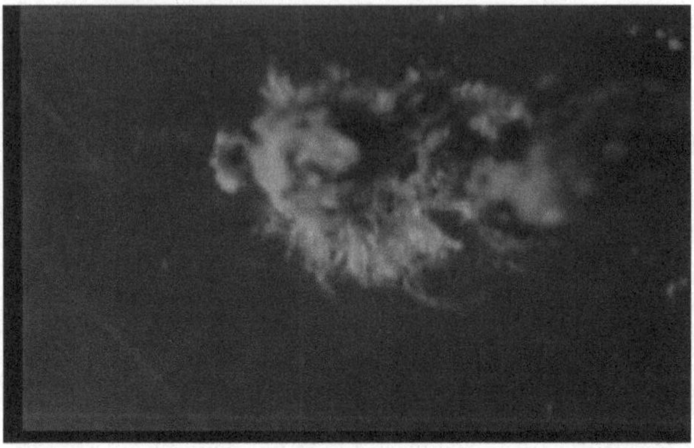

Fig. (29): 1st instar larva of *C. Vittata* infected with *M. anisopliae*, [12]

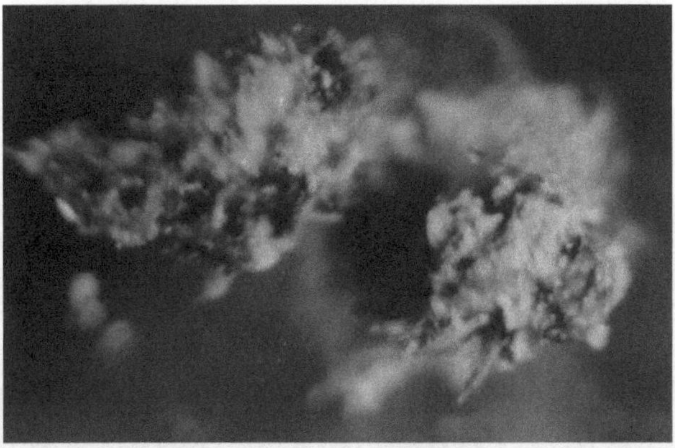

Fig. (30): 2nd instar larva of *C. vittata* infected with*M. anisopliae*, [12]

Fig. (31) : 3rd instar larva of *C. vittata* infected with *M. anisopliae*, [12]

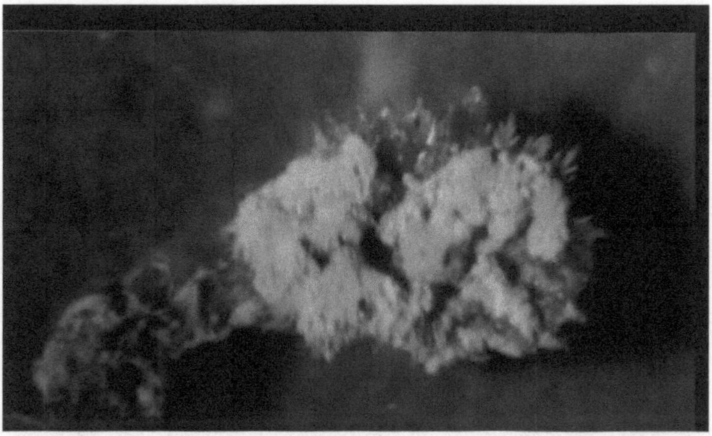

Fig. (32) : 4th instar larva of *C. vittata* infected with *M. anisopliae*, [12]

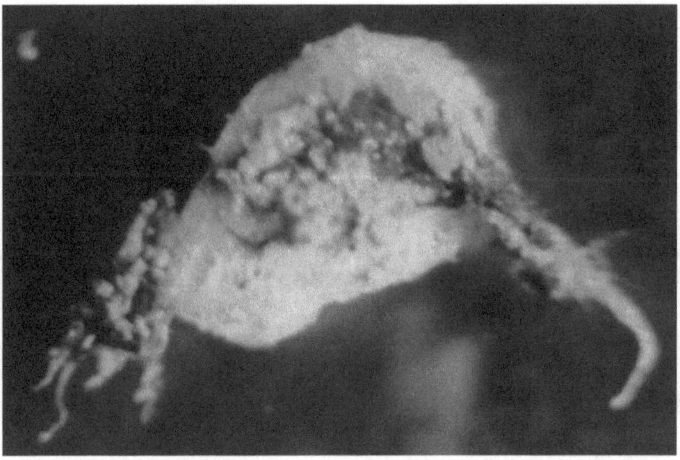

Fig. (33): 5th instar larva of *C. vittata* infected with *M. anisopliae*, [12]

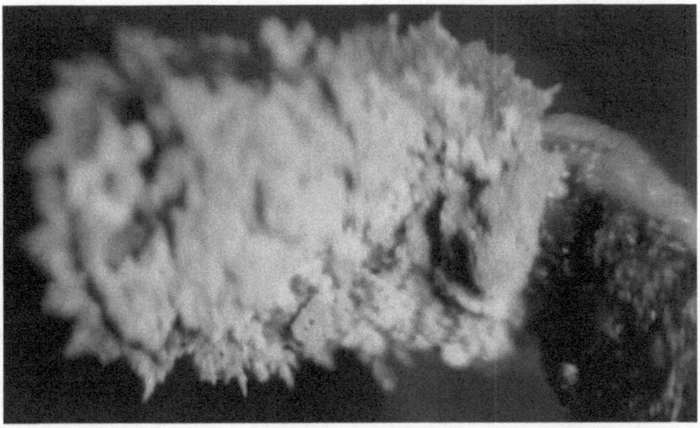

Fig. (34): Pre-pupa of C. *vittata* infected with *M. anisopliae*, [12]

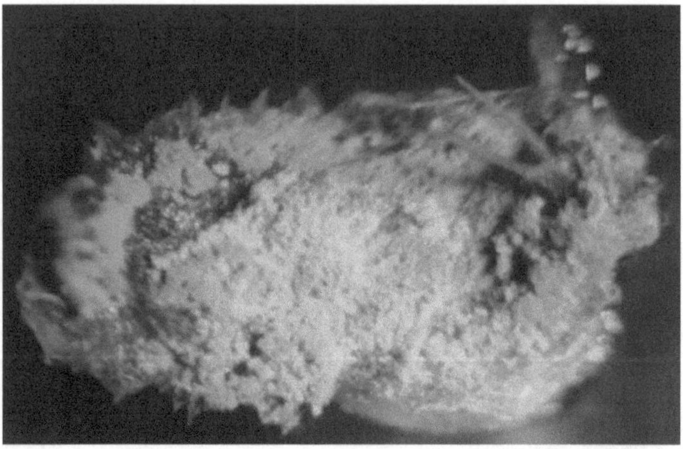

Fig. (35) Pupae of C. *vittata* infected with *M. anisopliae*, [12]

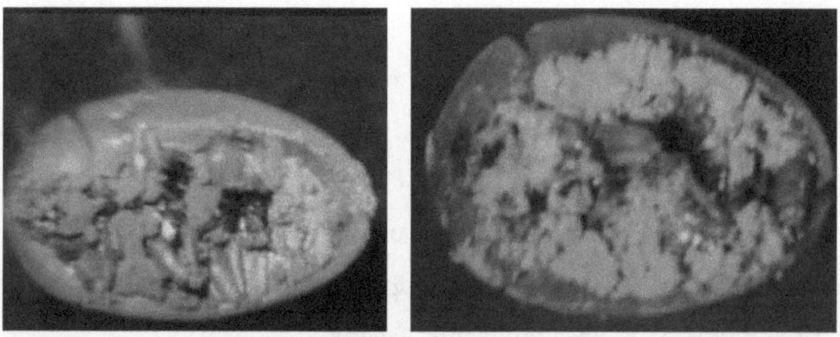

Fig. (36): Adult of *C. vittata* infected with *M. anisopliae*, [12]

Tomato leaf miner, *Tuta absoluta* (Meyrick)

Efficacy of 3 concentrations of *B. bassiana, M. anisopliae* and *V. lecanii* were prepared with concentrations of ($1x10^5$; $1x10^6$; $1x10^7$) were tested on *T. absoluta* larvae (Neonate "newly hatched", 2nd & 3rd instar), [51].

Eggs of *T. absoluta* were exposure to *B. bassiana, M. anisopliae* and *V. lecanii* to evaluate their effect on hatchability. The estimated LC_{50} of *V. lecanii* were 3.25 x 10^5 spores/ml, 5.47 x 10^5& 3.28 x $10^5$for neonate, 2nd instar & 3rd instar *T. absoluta* larvae, respectively. While the LC_{50} values of *B. bassiana* & *M. anisopliae* were (0.28 x 10^5 & 0.11 x 10^5), (0.45 x 10^5 & 0.46 x 10^5) and (0.32 x 10^5 & 0.27 x 10^5 conidia/ml) for neonate, 2nd instar & 3rd instar *T. absoluta* larvae, respectively. According to LC_{50} values, *B. bassiana* and *M. anisopliae* were most effective on larval phase of *T. absoluta* than *V. lecanii*, [51].

Daily mortality (%) of larval phase of *T. absoluta* fed in the newly hatched 2nd instar and 3rd instar larva with leaves treated with *V. lecanii*. The results revealed that when neonate larvae fed on *V. lecanii* the pathogen effect was evident by the third day of evaluation after exposure in the three concentrations (10^5; 10^6; 10^7spores/ml) with recorded mortality (13.3, 13.3, 33.3%) respectively, [52].

Thereafter, the values of the corrected mortalities of neonate larvae increased gradually from the 4th day after exposure until the last day (10th) with mortality 88.8% for the first concentration 10^5 spores/ml. For the second concentration (10^6 spores/ml) the mortality values were increased and reached its maximum in the 8th day of exposure (100% reduction). The mortality values reached its maximum in the 7th day and in the 5th day for the second and third concentration (10^6 & 10^7 spores/ml) and recorded 100% reduction for the two concentrations, [51].

The higher effective concentration of *V. lecanii* on neonate larvae of *T. absoluta* was 10^7spores/ml followed by 10^6 spores/ml while the other concentrations (10^5 spores/ml) showed moderate effect, [51].

For the second instar larva of *T. absoluta* the pathogen effect was evident by the fifth day for (10^5 spores/ml), by the 4th day for (10^6 spores/ml) and by the 3rd day for (10^7 spores/ml) of evaluation after exposure with corresponding mortalities (53.3, 40.0, and 6.6 %) for the three concentrations, respectively, [52]. The values of the corrected mortalities of 2nd instar larvae increased gradually until the 9th day after exposure to record 88.8% mortality for the first concentration (10^5 spores/ml), [51].

The mortality values reached its maximum in the 9th day, 8th day and in the 7th day for the three concentrations and recorded 100% reduction. Daily mortality (%) of larval phase of *T. absoluta* fed in the newly hatched "neonate", 2nd instar and 3rd instar larva with leaves treated with *B.bassiana*.

The higher effective concentration of *B. bassiana* on the 2nd instar larvae of *T. absoluta* resulted from the two concentrations 10^7 and 10^6conidia/ml evenly, followed by 10^7 and 10^6 conidia/ml, respectively, [51].

When 3rd instar larvae fed on *B. bassiana* the pathogen effect was evident by the 3rd day of evaluation after exposure in the first concentration (10^5 conidia/ml) with recorded mortality (20.0%).

While the pathogen effect was evident by the 2nd day of evaluation after exposure in the 1st, 2rd & 3th concentrations (10^5, 10^6 & 10^7) with recorded mortalities (6.7%) for the three concentrations, [51].

The potato tuber moth *Phthorimaea operculella* (ZELLER)

The LC_{50} of the target pest $1X10^6$ and $8.7\ X10^5$ spores /ml after *P. operculella* treated with *B. brongniartii* and *N. rileyi*, respectively. Under semi-field conditions, the corresponding LC_{50} recorded $1.20X10^6$ and $9.7\ X10^5$ spores /ml respectively), [52].

The effect of the entomopathogenic fungi against *P. operculella*, the number of eggs laid/ female recorded significant reduction to 42±1.7 and 33±8.9, after being treated with *B. brongniartii* and *N. rileyi* as compared to the control, respectively. The percentage of emerged adults was significantly decreased by (4%) as compared with the control, [52].

Cabbage Worm, *Pieris rapae* L.

Three isolates of *B. bassiana*, *M. anisopliae* and *V. lecanii* were evaluated against cabbage worm, *P. rapae* L. under laboratory and field conditions. *B. bassiana*, *M. anisopliae* and *V. lecanii* have not effect on cabbage worm, *Pieris rapae* L. after two days from treatment[53].

Mortalities are occurred in the 3rd day. The percent of mortalities are increased gradually and reached to the maximum in the 10th day from treatment.

The percent of mortalities ranged between 70 to 80, 60 to 75.7 and 40.0 to 55.2 % with *B. bassiana, M. anisopliae* and *V. lecanii*, respectively, in the 10th day after treatment. This means that *B. bassiana* isolation is more effective than *M. anisopliae* and *V. lecanii,* [53].

The % mortalities with all concentrations of *B. bassiana* isolation were 70, 72.4 and80 %, respectively.
Also, mortalities are occurred in the 3rd day. The percent of mortalities are increased gradually and reached to the maximum in the 10th day from treatment.

Data also showed a positive correlation between concentrations of fungi and the percentage of *P. rapae* L. Pupae mortality. The percent of mortalities ranged between 65 to 76, 58 to 70.1 and 42.0 to 60 % with *B. bassiana, M. anisopliae* and *V. lecanii*, respectively, in the 10th day after treatment. This means that *B. bassiana* isolation is more effective than *M. anisopliae* and *V. lecanii.* The percent of mortalities with all concentrations of *B. bassiana* isolation were 65, 70 and76 %, respectively [53].

Whitefly, *Bemisia tabaci* (Genn.)

There are no effect for *V. lecanii*, *M. anisopliae* and *B. bassiana* to *B. tabaci* after 3 days from treatment.

The percent of mortalities were increased gradually and reached to the maximum in the 7th day from treatment. With the all concentrations, the percent of mortalities were increased with increase of concentrations, [54].

The % mortalities ranged between 68.3 to 100, 56.2 to 90.5 and 62.7 to 97.3 % with *V. lecanii*, *M. anisopliae* and *B. bassiana*, respectively, in the 7th day after treatment. This mean that *V. lecanii* isolation is more effective than *M. anisopliae* and *B. bassiana*. The percent of mortalities with all concentrations of *V. lecanii* isolation were 68.3, 90.4and 100%, respectively, [54].

The number of *B. tabaci* per leave was decreased compared with control after the 2nd application. That the percent of reduction by *V. lecanii M. anisopliae* and *B. bassiana* after the third application were 57.3, 90.4 & 100, 50.2, 70.0 & 75.5 and 55.5, 79.5 & 90 % with C_1, C_2 and C_3, respectively[54].

There were no significant differences between the concentrations and controls after the first application all plots. After the 3rd application there were highly significant differences among all concentrations. The L.S.D was two, three after the third application. The statistical analysis confirmed that the third concentration (1 x 10^9) was the highly toxic compared the first and the second concentrations in all treatment, [54].

The percent of reduction by *V. lecanii M. anisopliae* and *B. bassiana* after the third application were 55.2, 85.2 & 100, 50.0, 71.0 & 76.2 and 56.2, 78.3 & 92 % with C_1, C_2 and C_3, respectively, [55].

The % of reduction by *V. lecanii M. anisopliae* and *B. bassiana* after the third application were 55.2, 85.2 & 100, 50.0, 71.0 & 76.2 and 56.2, 78.3 & 92 % with C_1, C_2 and C_3, respectively, [54].

Rhyzopertha dominica, _Sitophilus oryzae_ and _Oryzaephilus surinmensis._

R. dominica LC_{50} and LC_{90} values confirmed that *R. dominica* was more susceptible to *M. anisopliae* than *B. bassiana*, where LC_{50} and LC_{90} were 1.2 x 10 ^{5}, 2.3 x 10 ^{5} conidia /g and 2.7 x 10 ^{5,} 0.01 x 10 ^{6} conidia /g, respectively. The LT_{50}s were calculated as 4.6 and 6.6 days for the two respective fungi, indicating the superiority of *B. bassiana* over *M. anisopliae,* [53].

Sitophilus oryzae

The % cumulative mortality of *R. dominica* maximum mortality percentages adult were recorded 11 days after treatment with *B. bassiana* and *M. anisopliae*, where the mortality percentages were 25.2, 38.1, 45.2 and 50.3% for *B. bassiana* at the tested concentrations of 0.12 x 10 ^{6}conidia /g, 0.22 x 10 ^{6}conidia /g, 0.32 x 10 ^{6}conidia / g, and 0.42 x 10 ^{6} conidia /g, respectively, and 63.0, 72.3, 75.3, and 79.3% for *M. anisopliae* at the tested concentrations of 0.12 x 10 ^{6}conidia /g, 0.22 x 10 ^{6}conidia /g, 0.32 x 10 ^{6}conidia / g, and 0.42 x 10 ^{6} conidia / g, respectively.

The cumulative mortality percentages started with low levels on the 3rd day after treatment. Then the Percent mortality of *S. oryzae* was increased gradually, till the maximum was recorded on the 11th days of observation, where the mortality percentages were recorded 23, 32, 40 and 54% for *B. bassiana* respectively. The *S. oryzae* was more susceptible to infection with *B. bassiana* than *M. anisopliae.* LC_{50} and LC_{90} values proved that *S. oryzae* was more susceptible to *M. anisopliae* than *B. bassiana*, where the LC_{50} and LC_{90} of, *B. bassiana* and *M. anisopliae* were 4.3 x 10 ^{5} and 0.08 x 10 ^{6} conidia /g, and 1.6 x 10 ^{5} and 7.1 x 10 ^{5}conidia /g, respectively.

LT_{50}s were calculated as 8.8 and 6.3 days for the two respective fungi, indicating that the conidial spores of *B. bassiana* were less efficient on *S. oryzae* than those of *M. anisopliae,* [55].

Oryzaephilus surinmensis

The same effective was found when the entomopathogenic fungi were used against *O. surinmensis*. The concentrations 0.42 x 10^6 conidia / g of fungi, induced increase the mortalities rates.

The highest cumulative mortalities were recorded after 11 days of treatment when mortality percentages were recorded 25, 35, 43 and 51% for *B. bassiana* at the tested concentrations of (0.12 x 10^6conidia /g, 0.22 x 10^6conidia /g, 0.32 x 10^6conidia / g, and 0.42 x 10^6 conidia / g, respectively).

The mortality percentage of *O. surinmensis* treated with *B. bassiana* and *M. anisopliae* differed significantly among different concentrations. Also mortality percentage differed significantly between *O. surinmensis* treated with *B. bassiana* and with *M. anisopliae,* [55].

Application in the field of isolated fungi against Many Insects *S. ocellatella* and *C. Vittata*

The application of *B. bassiana* and *M. anisopliae* blast spores in sugar – beet fields was applied in Kafr El- Sheikh (Abu Kalab), Egypt. Sugar – beet plants were sprayed with the fungal suspensions to control *C. vittata*. Blastspores were grown on agitated bio malt liquid medium.

Blastspores of *B. bassiana* and *M. anisopliae* were applied to sugar- beet plants by using (10) lit. Knapsack sprayer with a concentration of (10^8 spores / ml.). Another untreated plot was used as control. In the treated and control plots, eggs, larvae, pupae and adults were counted on (25) plants .The spraying was repeated for three times in week, three weeks and five weeks, [60].

The field treatment of *C. vittata* with *M. anisopliae*

Using of *M. anisopliae* in the field application against the insect of *C. vittata*. Number of eggs, larvae, pupa and adult was counted and recorded in treatment and control before the beginning of the treatment and every two weeks after the application.

Before spraying the average number of eggs in treatment was 0.6 /plant and 0.26 / plant in control, the average number of larvae in treatment was 1.12/ plant and 1.2 / plant in control, the average number of pupa in treatment was 0 / plant and 0 / plant in control, the average number of adult in treatment was 0.6/ plant and 0.6 /plant.

After two weeks from the application the average number of eggs in treatment were 2.64/ plant and in control was 5.76 egg / plant, the average number of larvae in treatment was 3.32 /plant and in control was 5.24 / plant, average number of pupae in treatment was 2.0 / plant and in control was 2.08 /

plant, the average number of adult in treatment was 3.28 / plant and in control was 4.28 / plant.

After 4 weeks after treatment the average number of eggs in treatment was 1.84 / plant and in control was 3.44 / plant, the average number of larvae in treatment was 3.96 / plant and in control was 7.6 / plant, the average number of pupae in treatment was 1.12/plant and in control was 1.64/plant, the average number of adult in treatment was 2.36 / plant and in control was 4.84 /plant

After six weeks after treatment the average number of eggs was 4.84 / plant in treatment and 4.92 / plant in control, the average number of larvae was 5.52/ plant in treatment and 10.4 / plant in control, the average number of pupae was 0.92 / plant in treatment and 1.5/ plant in control, the average number of adult was 1.68 / plant in treatment and 2.92 /plant in control, [60].

Table (9) Average number of *C. vittata* different stages treatment with the entomopathogenic fungus *M. anisopliae,* [60]

Experimental Time	Average number of *C. vittata* / 25 plants							
	Eggs		Larvae		Pupae		Adults	
	T	C	T	C	T	C	T	C
Before First spray	0.6	0.68	1.12	1.2	0	0	0.6	0.6
Before Second spray	2.64	5.76	3.32	5.24	2.0	2.08	3.28	4.28
Before Third spray	1.84	3.44	3.96	7.6	1.12	1.64	2.36	4.84
After the Third spray	4.84	4.92	5.52	10.4	0.92	1.5	1.68	2.92

The field treatment of *C. vittata* with *B. bassiana:*

Using of *B. bassiana* in the field application against the insect of *C. vittata.* Number of eggs, larvae, pupa and adult was counted and recorded in treatment and control before the beginning of the treatment and every two weeks after the application.

Before spraying the average number of eggs in treatment was 6.0 /plant and 5.8 / plant in control, the average number of larvae in treatment was 4.8/ plant and 5.2 / plant in control, the average number of pupa in treatment was 1.9 / plant and 2.0 / plant in control, the average number of adult in treatment was 4.6/ plant and 4.2/plant in control, [60].

After two weeks from the application the average number of eggs in treatment were 1.88/ plant and in control was 3.44 egg / plant, the average number of larvae in treatment was 3.2 /plant and in control was 7.6 / plant, average number of pupae in treatment was 0.88 / plant and in control was 1.6 / plant, the average number of adult in treatment was 2.6 / plant and in control was 4.8/ plant, [60].

After four weeks after treatment the average number of eggs in treatment was 3.24 / plant and in control was 4.92/ plant, the average number of larvae in treatment was 4.2 / plant and in control was 10.4 / plant, the average number of pupae in treatment was 1.04/plant and in control was 1.5/plant, the average number of adult in treatment was 1.6 / plant and in control was 2.9 /plant, [60].

After 6 weeks after treatment the average number of eggs was 1.8 / plant in treatment and 3.64 / plant in control, the average number of larvae was 2.84/ plant in treatment and 7.7/ plant in control, the average number of pupae was 3.2 / plant in treatment and 7.2/ plant in control, the average number of adult was 2.7/ plant in treatment and 7.9/plant in control, [60].

Table (10): Average number of *C. vittata* different stages treatment with entomopathogenic fungus *B. bassiana*, [60]

Experimental Time	Average number of *C. vittata* / 25 plants							
	Eggs		Larvae		Pupae		Adults	
	T	C.	T	C	T	C	T	C
Before First Spray	6.0	5.8	4.8	5.2	1.9	2.0	4.6	4.2
Before 2nd spray	1.88	3.44	3.2	7.6	0.88	1.6	2.6	4.8
Before 3rd Spray	3.24	4.92	4.2	10.4	1.04	1.5	1.6	2.9
After 4th Spray	1.8	3.64	2.84	7.7	3.2	7.2	2.7	7.9

Table (11): Average number of *C. vittata* different stages treated with the entomopathogenic fungi *M. anisopliae,* [60]

Experimental Time	Average number of *C. vittata* / plants							
	Eggs		Larvae		Pupae		Adults	
	T	C	T	C	T	C	T	C
First week	0.6	0.68	1.12	1.2	0	0	0.6	0.6
Second week	2.64	5.76	3.32	5.24	2.0	2.08	3.28	4.28
Third week	1.84	3.44	3.96	7.6	1.12	1.64	2.36	4.84
Fourth week	4.84	4.92	5.52	10.4	0.92	1.5	1.68	2.92

Table (12): Average number of *C. vittata* different stages treated withentomopathogenic fungi *B. bassiana*, [60]

Experimental Time	Average number of *C. vittata* / plants							
	Eggs		Larvae		Pupae		Adults	
	T	C.	T	C	T	C	T	C
First week	6	5.8	4.8	5.2	1.9	2.0	4.6	4.2
Second week	1.88	3.44	3.2	7.6	0.88	1.6	2.6	4.8
Third week	3.24	4.92	4.2	10.4	1.04	1.5	1.6	2.9
Fourth week	1.8	3.64	2.84	7.7	3.2	7.2	2.7	7.9

Table (13) Average number of *C. vittata* different stages treatment with the entomopathogenic fungi *M. anisopliae,* [60]

Experimental Time	Average number of *C. vittata* / 25 plants							
	Eggs		Larvae		Pupae		Adults	
	T	C	T	C	T	C	T	C
First week	0.6	0.68	1.12	1.2	0	0	0.6	0.6
Second week	2.64	5.76	3.32	5.24	2.0	2.08	3.28	4.28
Third week	1.84	3.44	3.96	7.6	1.12	1.64	2.36	4.84
Fourth week	4.84	4.92	5.52	10.4	0.92	1.5	1.68	2.92

Table (14): Average number of *C. vittata* different stages treatment with entomopathogenic fungi *B. bassiana,* [60]

Experimental Time	Average number of *C. vittata* / 25 plants							
	Eggs		Larvae		Pupae		Adults	
	T	C.	T	C	T	C	T	C
First week	6	5.8	4.8	5.2	1.9	2.0	4.6	4.2
Second week	1.88	3.44	3.2	7.6	0.88	1.6	2.6	4.8
Third week	3.24	4.92	4.2	10.4	1.04	1.5	1.6	2.9
Fourth week	1.8	3.64	2.84	7.7	3.2	7.2	2.7	7.9

Table (15): Efficacy of the Fungi isolations against C. *vittata* larvae in sugar beet field, [60].

Fungi isolation	Conc. (ml)	Mean No. of alive larvae / plant & % Reduction after 4th applications during four weeks									Avg. % Red.
		1st spray			2nd spray		3rd spray		4th spray		
		Population before 1st spray	Population after one week	% Red.	Population after two week From spray	% Red.	Population after three weeks	% Red.	Population after four weeks	% Red.	
B. bassiana (B_6As)	0.5	4.0a	3.0b	36.0	2.2b	64.7	2.0b	52.7	1.6b	84.0	59.35
	1.0	3.8a	2.2bcd	51.0	1.6b	73.0	1.2bc	83.8	1.0b	88.2	74.0
	2.0	3.8a	1.4cde	68.5	1.0b	83.1	0.8c	89.8	0.6b	92.9	83.6
M. anisopliae (M_4As)	0.5	4.0a	2.4bc	48.6	2.0b	67.9	1.4bc	82.0	1.2b	86.5	71.3
	1.0	4.0a	1.2de	74.3	1.2b	80.8	0.8c	89.8	0.4b	95.5	85.1
	2.0	3.8a	0.8e	82.0	0.8b	86.5	0.4c	94.6	0.2b	97.3	90.1
control	0.0	3.6a	4.2a	----	5.6a	----	7.0a	-----	8.0a	-----	------

				-		-					
F		0.02 un	8.9*		10.87*		38.9*		21.5*		------
$LSD_{0.05}$		2.65	1.13		1.44		1.65		1.7		------

Treatment means followed by the same letter are not significantly different from each other (P < 0.05).

Table (16): Efficacy of the Fungi isolations against C. *vittata* larvae in sugar beet field,[60].

Fungi isolation	Conc. (ml)	Mean No. of alive larvae / plant & % Reduction after 4th applications during four weeks									Avg. % Red.
		1st spray			2nd spray		3rd spray		4th spray		
		Population before 1st spray	Population after one week	% Red.	Population after two week From spray	% Red.	Population after 3 weeks	% Red.	Population after 4weeks	% Red.	
B. bassiana (B6As)	0.5	3.0a	2.6ab	41.0	2.0b	56.7	1.6b	69.9	1.0b	85.1	63.2
	1.0	3.0a	2.0ab	55.0	1.6bc	65.4	0.6c	88.7	0.6b	91.1	75.8
	2.0	3.0a	1.6ab	64.0	1.0bc	78.4	0.4c	92.5	0.4b	94.1	82.25
M. anisopliae (M4As)	0.5	3.0a	2.0ab	54.4	1.0bc	78.4	0.6c	88.7	0.8b	88.1	77.4
	1.0	3.0a	1.6ab	63.6	0.8c	82.7	0.6c	88.7	0.4b	94.1	82.3
	2.0	2.6a	1.2ab	68.5	0.6c	85.7	0.4bc	91.4	0.2b	96.6	85.6

control	0.0	2.6a	3.8a	---- -	4.0a	-----	4.6a	----	5.8a		----
F		0.08un	2.96un		10.05*		19.7*		18.42*		
$LSD_{0.05}$		1.93	1.45		1.07		0.17		1.09		

Treatment means followed by the same letter are not significantly different from each other (P < 0.05)

The Potato tuber moth, *Phathorimaea operculella* (ZELLER)

Application of the fungi occurred was accomplished at the rate of 1X 10^8 spores/ml. Applications were made by a sprayed at the sunset. Four applications were made at 4- weeks intervals during crop growing season. Control plots were left without any treatments. Examinations of 40 plants /plot /treatment were carried out just before the first application and seven days after last application to calculate the average reduction percentages in the target insect infestation percentages which were calculated in each treatment. The agricultural practices followed the recommendations of the Ministry of Agricultural. Twenty tubers were taken from the first 5 rows in each treatment and in the control, [53].

Under field conditions the yields weight of potatoes were significantly increased to 25.45± 55.66 and 29.67±61.11 ton/ feddan in plots treated with *B. brongniartii* and *N. rileyi* respectively as compared to17.88±55.43 Tons/ feddan in the control. The weight of potatoes were significantly increased to 25.97±67.91 and 29.94±54.98 Ton/ feddan as compared to 12.27±45.09 in the control, [53].

Sitophilus granarius

The effect of the grain weevil *S. granarius* were significantly decreased after oil extract treatments , the percentage of larval mortality , 88.14, 39.13 and 2.21 after *N. Sativa* treatments at the concentrations of 5, 0.5 and 0.05%. The corresponding concentrations of *T. distichum* gave the larval mortality 57.18, 20.26 and 4.41 % of the *S. granarius,* respectively. When *B. carterii* treated with the last concentrations the larval mortality were significant decrease to 29.61, 15.81 and 0.613% as compared to zero in the control, [55].

The oviposition deterrent effect in the means number of eggs were significantly decreased to 13.0±7.8, 37±7.9 and 45±3.8 eggs/ female as when treated with 3% of the oil *N. sativa , T. distichum and B. carterii*, respectively compared to 198±4.9 in the control,[55].

The half-life period of the target insect pest after oil extract treatments which show that the LC_{50} obtained 2.1 X 10^{5}, 3.2 X 10^{5}, and 3.7 X 10^{5} as compared to 8.8 X 10^{5} sores /ml. [55].

Accumulative mortality of *S. granarius* during the first week of rice seeds exposed to treated foam with *N. sativa, T. distichum,* and *B. carterii* on *S. granarius* oils. After seven days from treated rice seeds the accumulative mortality was 78.15, 50.17, and 30.1, respectively compared with 2.1 in untreated, [55].

Pieris rapae

There are significant difference between 1st (C_1) and 2nd (C_2) spores concentrations and control after the first application in all parts, the differences appear gradually after the second and third application. On the other hand the third concentration (C_3) in *B. bassiana* was the best concentration against *P. rapae* L. followed by the third concentration in *M. anisopliae* and the third concentration in *V. lecanii*. This means that the third concentration in *B. Bassiana* was the best concentrations against *P. rapae,* [53].

There are significant difference also between 1st (C_1) and 3rd (C_2) spores concentrations and control after the first application in all parts, the differences appear gradually after the second and third application. On the other hand the third concentration (C_3) in *B. bassiana* was the best concentration against *P. rapae* followed by the third concentration in *M. anisopliae* and the third concentration in *V. lecanii*. The Percent of reduction in all treatment was ranged between 48.5 to 75.8% in all concentrations, [53].

B. Bassiana, *M. anisopliae* and *V. lecanii* isolates are promising agents for Cabbage worm control in the field, [53].

Sugar beet fly, *Pegomyia mixta*

P. mixta started to appear during the second week of January. In mid-February, the first application was carried out using two entomopathogenic fungi *B. bassiana,* and *M. anisopliae,* [57].

The number of infested / 10 plants was 70 % before the first spray by *B. bassiana* and reached to 10 % after seven days from treatment and reached to zero% infested after 14 day from first treatment. Also, the number of infested leaves 11 before the first spray by *B. bassiana* and reached to 4 after seven days from treatment and reached to zero infested after 14 day from first treatment, [57]. The number of infested egg patches / plant 13 before the first spray by *B. bassiana* and reached to 5 after seven days from treatment and reached to zero infested after 14 day from first treatment.

The number of infested larvae / plant 22 before the first sprays by *B. bassiana* and reached to 4 after seven days from treatment and reached to zero infested after 14 day from first treatment.

The number of infested / 10 plants was 80 % before the first spray by *M. anisopliae* and reached to 20 % after seven days from treatment and reached to 10% infested after 14 day from first treatment. Also, the number of infested leaves 12 before the first spray by *B. bassiana* and reached to 6 after seven days from treatment and reached to 2% infested after 14 day from first treatment.

The number of infested egg patch / plant 12 before the first spray by *B. bassiana* and reached to 6 after seven days from treatment and reached to 3% infested after 14 day from first treatment.

The number of infested larvae / plant 21 before the first spray by *B. bassiana* and reached to 15 after seven days from treatment and reached to zero infested after 14 day from first treatment, [57].

The number of infested / 10 plants 60 % before the first spray by *B. bassiana* and reached to 20 % after seven days from treatment and reached to zero% infested after 14 day from first treatment. Also, the number of infested leaves 13 before the first spray by *B. bassiana* and reached to 3 after seven days from treatment and reached to zero infested after 14 day from first treatment.

The number of infested egg patches / plant 12 before the first spray by *B. bassiana* and reached to 4 after seven days from treatment and reached to zero infested after 14 day from first treatment, [57].

The number of infested larvae / plant 25 before the first spray by *B. bassiana* and reached to 12 after seven days from treatment and reached to zero infested after 14 day from first treatment.

The number of infested / 10 plants 70 % before the first spray by *Metarhizium anisopliae* and reached to 30 % after seven days from treatment and reached to Zero% infested after 14 day from first treatment. Also, the number of infested leaves 13 before the first spray by *Metarhizium anisopliae* and reached to 6 after seven days from treatment and reached to Zero% infested after 14 day from first treatment.

The number of infested egg patches / plant 12 before the first spray by *M. anisopliae* and reached to 6 after seven days from treatment and reached to Zero% infested after 14 day from first treatment.

The number of infested larvae / plant 23 before the first sprays by *M. anisopliae* and reached to 14 after seven days from treatment and reached to zero infested after 14 day from first treatment.

The entomopathogenic fungi *B. bassiana* and *M. anisopliae* can be used as a promising agent in pest control and integrated pest management programs instead of conventional pesticides to reduce the environmental pollution especially when the pests were under the economic threshold, [57].

The Oliver black scale insect, *Saissetia oleae* (Oliver)

Percent of sampled substrate infested with black scale Insect, *S. oleae* (Olivier) before and after treatment with *B. bassiana* reduction from 92.0% in the first week from treatment to 9.2 % after seven weeks from treatment, [58].

Also when we treated the trees with *M. anisoplia*e reduction in population of the black scale Insect, *S. oleae* was 2.1% from 93.0% after seven weeks from treatment, [58].

Also when we treated the trees with *V. lecanii* reduction in population of the black scale Insect, *S. oleae* was 9.5% from 90.0% after seven weeks from treatment. *M. anisoplia*e caused the highest reduction in black scale Insect, *S. oleae* than *B. bassiana* and *V. lecanii*, [58].

Percent of sampled substrate infested with black scale Insect, *S. oleae* (Olivier) before and after treatment with *B. bassiana* reduction from 84.0% in the first week from treatment to 8.3 % after seven weeks from treatment, , [58].

Also when we treated the trees with *M. anisoplia*e reduction in population of the black scale Insect, *S. oleae* was 1.0 % from 86.0% after seven weeks from treatment.

Also when we treated the trees with *V. lecanii* reduction in population of the black scale Insect, *S. oleae* was 8.5% from 85.0% after seven weeks from treatment. *M. anisoplia*e caused the highest reduction in black scale Insect, *Saissetia oleae* than *B. bassiana* and *V. lecanii*, [58].

Anti-soft scale insect *Pulvinaria tenuivalvata*

The mean numbers of adults found on leaves before treatments, ranged from 58.6 to 68.2scales / 50 leaves, indicating a relatively uniform distribution of insect infestation. Three days after spraying, the treatments suppressed the levels of infestation to different degrees as compared to untreated control. *Beauveria bassiana* (10^8 spores/ml), Petroleum ether fraction of *Cressa cretica* with a rate of (50ml/L.) and significantly lowered the percentage of infestation to 39.9% and 44.6 %, respectively, although they didn't reach KZ oil (15ml/L.) activity (54.0%). Two weeks after treatments, *B. bassiana* (10^8 spores/ml), Petroleum ether fraction (50ml/L.) and KZ oil (15ml/L.) had almost similar activity, 65.8,69.1 and 79.4% reduction in the infestation, respectively.

Statistical analysis showed significant differences between tested compounds and control on the soft scale insect population. The low susceptibility of the adult stage to the tested compounds may be attributed to the presence of protective scales which prevent the penetration of sprayings, [59].

Increased percentages over the control ranged between 1.23 to 1.27 when used *Beauveria bassiana*, *Cressa cretica* Petroleum ether fraction 5% and KZ oil 95%, respectively. The average calculated yield in the control was 21710 Kg / feddan, as compared to 26750, 27180 and 27670 Kg / feddan after using the *Beauveria bassiana*, *Cressa cretica* Petroleum ether fraction 5% and KZ oil 95%, respectively, [59].

The green stink bug, *Nezara viridula* L. in sugar-beet

Table 6 showed that the third concentration (3×10^5) in *Metarhizium anisopliae* was the best concentration against *Nezara viridula* followed by the third concentration in *B. bassiana* isolation. The same results were found by El-Zoghby (2003) and Abdel-Raheem (2009 and 2010).

After the second application the average number of *Nezara viridula* decreased compared with control. That the average numbers by *Metarhizium anisopliae* after the third applications were 2.0, 1.0 and 0.0 with 3×10^3, 3×10^4 and 3×10^5, respectively and the average numbers by *B. bassiana* after the third applications were 4.0, 2.0 and 1.0 with 3×10^3, 3×10^4 and 3×10^5, respectively, **[61]**

During 2010 season Table 7 showed that the third concentration (3×10^5) in *Metarhizium anisopliae* was the best concentration against *Nezara viridula* followed by the third concentration in *B. bassiana* isolation. The same results were found by El-Zoghby (2003) and Abdel-Raheem (2009 and 2010).

After the second application the average number of *Nezara viridula* decreased compared with control. That the average numbers by *Metarhizium anisopliae* after the third applications were 1.0, 1.0 and 0.0 with 3×10^3, 3×10^4 and $3x\ 10^5$, respectively and the average numbers by *B. bassiana* after the third applications were 7.0, 3.0 and 2.0 with 3×10^3, 3×10^4 and 3×10^5, respectively, **[61]**

Finally, these data clear that the entomopathogenic fungi *Metarhizium anisopliae* and *B. bassiana* can be used as a promising agent in pest control and integrated pest management programs instead of conventional pesticides to reduce the environmental pollution especially when the pests were under the economic threshold. The entomopathogenic fungi *Metarhizium anisopliae* was more promising than *B. bassiana* on *Nezara viridula*, **[61]**

Table (17): Average number of *Nezara viridula* on sugar-beet plants before spraying the two entomopathogenic fungi during Season (2009), [61]

Before Spray	Cont.	*M. anisopliae*			*B. bassiana*		
		3×10^3	3×10^4	$3x\ 10^5$	3×10^3	3×10^4	$3x\ 10^5$
1st	9	8	9	9	10	8	7
2nd	12	5	5	4	9	7	5
3rd	10	3	2	2	7	6	3
4th	5	2	1	0	4	2	1

Table (18): Average number of *Nezara viridula* on sugar-beet plants before spraying the two entomopathogenic fungi during season (2010), [61]

:

Before Spray	Cont.	*M. anisopliae*			*B. bassiana*		
		3×10^{3}	3×10^{4}	3×10^{5}	3×10^{3}	3×10^{4}	3×10^{5}
1st	13	10	12	10	12	13	11
2nd	10	7	7	5	10	9	7
3rd	14	3	4	2	9	6	4
4th	7	1	1	0	7	3	2

Application of Entomopathogenic Fungi as Nano

The Red Palm Weevil, *Rhynchophorus ferrugineus* Olivier (Coleoptera: Curculionidae)

Data of the treated eggs of *R. ferrugineus* with *M. anisopliae*, *B. bassiana*, Bio Magic, Bio Power and Bio Catch as fungal spores and their Silver NPs particles was presented in table (19). Seven days post treatment, up- to (90%) mortality rate of *R. ferrugineus* was recorded in the treated eggs.[62]

The % mortality rates attained 80, 73, 65, 60, and 45 % by infection with fungal spores from *M. anisopliae*, *B. bassiana*, Bio Magic, Bio Power and Bio Catch, respectively.

Furthermore, the % mortality rates were 90, 84, 73, 70, and 58% by infection with bio synthesized Ag NPs from *M. anisopliae*, *B. bassiana*, Bio Magic, Bio Power and Bio Catch, respectively during the same time. *M. anisopliae* recorded the highest mortality (90%) in the eggs of *R. ferrugineus,* when treated with Nano-particles or with fungal spores after 6 days and was the lowest (58%) when treated with Bio Catch. According to Abdel-Raheem *et al.* [51] the total mortality of eggs and larvae were reduced than the control group when eggs were exposed to *M. anisopliae* spores.

(Abdel-Raheem, [62]) mentioned that the reason of different pathogenicity rates between one fungus and another may due to the fraction exhibiting antimicrobial activity of some polar compounds ranging between 1000 and 1500 Dalton in the extraction of fungi.

Table (19): Mortality percent of entomopathogenic fungi on the eggs of *Rhynchophorus ferrugineus,* using spore suspension and bio synthesized silver nanoparticles [62]

Entomopathogenic fungus	Treated with	
	Fungal spores (Mean ± S.E)	Bio synthesized Ag NPs (Mean ± S.E)
Metarhizium anisopliae	80.0 ± 1.20	90.0 ± 2.10
Beauveria bassiana	73.0 ± 1.00	84.0 ± 1.30
Bio Magic	65.0 ± 0.20	73.0 ± 0.10
Bio Power	60.0 ± 3.10	70.0 ± 0.20
Bio Catch	45.0 ± 0.20	58.0 ± 1.20
Control	6.0	7.0
S.E (m)	1.14	1.98

Data of larval mortality of *R. ferrugineus,* treated with *M. anisopliae*, *B. bassiana*, Bio Magic, Bio Power and Bio Catch as fungal Spores and Silver Nano-particles was presented in table (20). Seven days post treatment, 90% mortality of *R. ferrugineus* larvae was recorded as: 84, 75, 71, 65, and 55 % by infection with fungal spores from *M. anisopliae*, *B. bassiana*, Bio Magic, *Bio* Power and Bio Catch, respectively. By infection with bio synthesized Ag NPs from *M. anisopliae*, *B. bassiana*, Bio Magic, Bio Power and Bio Catch, the mortality rates recorded were 95, 87, 77, 73, and 60 %, respectively. *M. anisopliae* was the highest % mortality (95%) in the larvae of *R. ferrugineus,* when treated with the Nano-particles or with the fungal spores after 6 days, while the lowest (60%) was when treated with Bio Catch, *V. lecanii*. The results agree with Tefera and Pringle (2003) who reported that the bio efficacy of *M. anisopliae* in all stages of *R. ferrugineus* caused up to 48 to 95 % mortality of adult and larvae. [62]

Table (20): Mortality percent of entomopathogenic fungi on the larvae of *Rhynchophorus ferrugineus* using spore suspension and bio synthesized silver nanoparticles [62]

Entomopathogenic fungus	Treated with	
	Fungal spores (Mean ± S.E)	Bio synthesized Ag NPs (Mean ± S.E)
Metarhizium anisopliae	84.0 ± 2.10	95.0 ± 2.30
Beauveria bassiana	75.0 ± 2.12	87.0 ± 2.22
Bio Magic	71.0 ± 2.20	77.0 ± 1.10
Bio Power	65.0 ± 1.10	73.0 ± 1.20
Bio Catch	55.0 ± 0.10	60.0 ± 1.00
Control	6.2	7.3
S.E (m)	2.10	3.00

Data of adult mortality of *R. ferrugineus,* treated with *M. anisopliae*, *B. bassiana*, Bio Magic, Bio Power and Bio Catch as fungal Spores and Silver Nano-particles was presented in table (21). Seven days post treatment, the parentage of mortality of *R. ferrugineus* adults reached 77%. The % mortality rates recorded 65, 61, 56, 52, and 35 % by infection with fungal spores from *M. anisopliae*, *B. bassiana*, Bio Magic, *Bio* Power and Bio Catch, respectively. While, by infection with bio synthesized Ag NPs from *M. anisopliae*, *B. bassiana*, Bio Magic, Bio Power and Bio Catch, the % mortality recorded were 77, 70, 63, 55, and 48 %, respectively. *M. anisopliae* was the highest % mortality (77%) in the adults of *R. ferrugineus,* when treated with the Nano-particles or with fungal spores after 6 days, while the lowest was (48%), when they were treated by Bio Catch.

Table (21): Mortality percent of entomopathogenic fungi on the adults of *Rhynchophorus ferrugineus* using spore suspension and bio synthesized silver nanoparticles [62]

Entomopathogenic fungus	Treated with	
	Fungal spores (Mean ± S.E)	Bio synthesized Ag NPs (Mean ± S.E)
Metarhizium anisopliae	65.0 ± 2.00	77.0 ± 1.30
Beauveria bassiana	61.0 ± 1.20	70.0 ± 2.10
Bio Magic	56.0 ± 2.20	63.0 ± 0.10
Bio Power	52.0 ± 1.20	55.0 ± 3.20
Bio Catch	35.0 ± 0.00	48.0 ± 0.20
Control	7.0	8.0
S.E(m)	1.94	2.90

The Ag NPs synthesized through application of spore suspension and bio synthesized silver Nano-particles showed an efficacy against *R. ferrugineus* different stages. *M. anisopliae* had the highest potential and was more effective than all the others.

Virulence of Nano – Particle preparation of Entomopathogenic fungi and Entomopathogenic Bacteria against red palm weevil *Rhynchophorus ferrugineus* (Olivier) (Coleoptera: Curculionidae).

Rhynchophorus ferrugineus is one of the most severe pests on palm species, including date palms In Asia. The purpose of the current study is to evaluate the virulence of bio efficacy of nano - particle of entomopathogenic fungi and entomopathogenic bacteria on red palm weevil, *Rhynchophorus ferrugineus.* Prepare concentrations of Fungal Spores and Silver Nano Particles were prepared from *Metarhizium anisopliae*, *Beauveria bassiana, Verticillium lecanii* and *B. thuringiensis*, *B. subtilis.* [63]

M. anisopliae was the highest % mortality (90%) ,(95%) and (77 %) against eggs, larvae and adults of *R. ferrugineus* , respectively when treated with bio synthesized or with fungal spores after seven days from treatment and the lowest % mortality was recorded when treated with *V. lecanii.* [63]

Three concentrations, (10^3, 10^4, and 10^5 CFU/mL), formulated as bacterial suspensions and bio-synthesized silver nanoparticles, (Ag NPs) were evaluated for their ability to inhibit egg hatch. The 10^5 CFU/mL concentration, both as a bacterial suspension and formulated as Ag NPs, was tested for efficacy against 10 day-old larvae and adults. [63]

Egg hatch was significantly inhibited by all concentrations and formulations of *B. subtilis* and *B. thuringiensis*, and exhibited lethal concentrations for 50% mortality (LC_{50}) of 3.45 X 10^3 and 6.73 X 10^4 CFU/mL, respectively. The percent mortality of larvae six days after treatment was 85% and 77% for *B. thuringiensis* and *B. subtilis* Ag NPs, respectively. The percent mortality of adults six days after treatment was 75% and 67% for *B. thuringiensis* and *B. subtilis* Ag NPs, respectively. [63]

The efficacy of nanoparticle preparations of *Metarhizium anisopliae*, *Beauveria bassiana,* and *Verticillium lecanii* against eggs, larvae, and adults of the red palm weevil. [63]

Table (22 shows that the results of the eggs of *R. ferrugineus* which were treated with *M. anisopliae*, *B. bassiana*, Bio Magic, Bio Power and Bio Catch as fungal Spores and Silver nano particles. After six days of treatment, the petri dishes plates were observed and the result was 90% mortality of *R. ferrugineus* eggs. Additionally, white, green and metallic muscardine were found on the dead eggs. The % mortality after six days, from treated subjects, recorded 80, 73, 65, 60, and 45 % by infection with fungal spores from *M. anisopliae*, *B. bassiana*, Bio Magic, Bio Power and Bio Catch, respectively. [63]

Furthermore, the % mortality after six days from treated subjects recorded 90, 84, 73, 70, and 58 % by infection with bio synthesized Ag NPs from *M. anisopliae*, *B. bassiana*, Bio Magic, Bio Power and Bio Catch, respectively. *M. anisopliae* was the highest % mortality (90%) on the eggs of *R. ferrugineus* when treated with bio synthesized or with fungal spores after six days from treatment and the lowest % mortality was (58%) when treated with Bio Catch. [63]

Table (22): Bio efficacy of entomopathogenic fungi on the eggs of *R. ferrugineus* using spore suspension and bio synthesized silver nanoparticles.[63]

Entomopathogenic fungi	Treated with	
	Fungal spores (Mean ± S.E)	Bio synthesized Ag NPs (Mean ± S.E)
M. anisopliae	80.0 ± 1.20	90.0 ± 2.10
B. bassiana	73.0 ± 1.00	84.0 ± 1.30
Bio Magic	65.0 ± 0.20	73.0 ± 0.10
Bio Power	60.0 ± 3.10	70.0 ± 0.20
Bio Catch	45.0 ± 0.20	58.0 ± 1.20
Control	6.0	7.0
S.E(m)	1.14	1.98

Table (23) shows that the 4th larvae of *R. ferrugineus* were treated with *M. anisopliae*, *B. bassiana*, Bio Magic, Bio Power and Bio Catch as fungal Spores and Silver nano particles. After six days of treatment, the petri dishes plates showed up to 90% mortality of *R. ferrugineus* larvae. [63]

Also, white, green and metallic muscardine were found on the dead larvae. The % mortality after six days recorded 84, 75, 71, 65, and 55 % by infection with fungal spores from *M. anisopliae*, *B. bassiana*, Bio Magic, *Bio* Power and Bio Catch, respectively. Furthermore, the % mortality after six days recorded 95, 87, 77, 73, and 60 % by infection with bio synthesized Ag NPs from *M. anisopliae*, *B. bassiana*, Bio Magic, Bio Power and Bio Catch, respectively. *M. anisopliae* was the highest % mortality (95%) against the larvae of *R. ferrugineus* when treated with bio synthesized or with fungal spores after six days and the lowest % mortality was (60%) when treated with Bio Catch, *V. lecanii*. [63]

Table (23): Bio efficacy of entomopathogenic fungi on the 4th larvae of *R. ferrugineus* using spore suspension and bio synthesized silver nanoparticles.

Entomopathogenic fungi	Treated with	
	Fungal spores (Mean ± S.E)	Bio synthesized Ag NPs (Mean ± S.E)
M. anisopliae	84.0 ± 2.10	95.0 ± 2.30
B. bassiana	75.0 ± 2.12	87.0 ± 2.22
Bio Magic	71.0 ± 2.20	77.0 ± 1.10
Bio Power	65.0 ± 1.10	73.0 ± 1.20
Bio Catch	55.0 ± 0.10	60.0 ± 1.00
Control	6.2	7.3
S.E(m)	2.10	3.00

Table (24) shows that the adults of *R. ferrugineus* were treated with *M. anisopliae*, *B. bassiana*, Bio Magic, Bio Power and Bio Catch as fungal Spores and Silver nano particles.[63]

After six days of treatment, the petri dishes plates were observed and the parentage of mortality of *R. ferrugineus* adults was 77%. Also, white, green and metallic muscardine were found on the dead adults. The % mortality after six days recorded 65, 61, 56, 52, and 35 % by infection with fungal spores from *M. anisopliae*, *B. bassiana*, Bio Magic, Bio Power and Bio Catch, respectively. Moreover, the % mortality after six days of treatment recorded 77, 70, 63, 55, and 48 % by infection with bio synthesized Ag NPs from *M. anisopliae*, *B. bassiana*, Bio Magic, Bio Power and Bio Catch, respectively. *M. anisopliae* was the highest % mortality (77%) against the adults of *R. ferrugineus* when treated with bio synthesized or with fungal spores after six days and the lowest % mortality was (48%) when it treated with Bio Catch[63]

Table (24): Bio efficacy of entomopathogenic fungi on the adults of *R. ferrugineus* using spore suspension and bio synthesized silver nanoparticles. [63]

Entomopathogenic fungi	Treated with	
	Fungal spores (Mean ± S.E)	Bio synthesized Ag NPs (Mean ± S.E)
M. anisopliae	65.0 ± 2.00	77.0 ± 1.30
B. bassiana	61.0 ± 1.20	70.0 ± 2.10
Bio Magic	56.0 ± 2.20	63.0 ± 0.10
Bio Power	52.0 ± 1.20	55.0 ± 3.20
Bio Catch	35.0 ± 0.00	48.0 ± 0.20
Control	7.0	8.0
S.E(m)	1.94	2.90

Silver Nano-particles from Entomopathogenic Fungi on The tortoise beetle, *Cassida vittata* (Coleoptera: Chrysomelidae)

Data of the treated eggs of *C. vittata* with *M. anisopliae*, *B. bassiana*, Bio Magic, Bio Power and Bio Catch as fungal spores and their Silver NPs particles was presented in table (25).[64]

Seven days post treatment, up- to (95%) mortality rate of *C. vittata* was recorded in the treated eggs. The % mortality rates attained 84, 75, 68, 65, and 47 % by infection with fungal spores from *M. anisopliae*, *B. bassiana*, Bio Magic, Bio Power and Bio Catch, respectively. Furthermore, the % mortality rates were 95, 87, 75, 72, and 61% by infection with bio synthesized Ag NPs from *M. anisopliae*, *B. bassiana*, Bio Magic, Bio Power and Bio Catch, respectively during the same time. *M. anisopliae* recorded the highest mortality (95%) in the eggs of *C. vittata,* when treated with Nano-particles or with fungal spores after 6 days and was the lowest (61%) when treated with Bio Catch. According to [12, 51, 61-64] the total mortality of eggs and larvae were reduced than the control group when eggs were exposed to *M. anisopliae* spores.[64]

Table (25): % Mortality of EPF on the eggs of *C. vittata,* using spore suspension and bio synthesized silver nanoparticles [64]

EPF	Treated with	
	Fungal spores (Mean ± S.E)	Bio synthesized Ag NPs (Mean ± S.E)
M. anisopliae	84.0 ± 1.22	95.0 ± 2.11
B. bassiana	75.0 ± 1.10	87.0 ± 1.20
Bio Magic	68.0 ± 0.10	75.0 ± 0.11
Bio Power	65.0 ± 2.13	72.0 ± 1.11
Bio Catch	47.0 ± 1.12	61.0 ± 1.25
Control	5.0	5.0
S.E (m)	1.13	1.16

Data of larval mortality of *C. vittata,* treated with *M. anisopliae*, *B. bassiana*, Bio Magic, Bio Power and Bio Catch as fungal Spores and Silver Nano-particles was presented in table (26).

Seven days post treatment, 97% mortality of *C. vittata* larvae was recorded as: 85, 78, 72, 67, and 58 % by infection with fungal spores from *M. anisopliae*, *B. bassiana*, Bio Magic, *Bio* Power and Bio Catch, respectively.

By infection with bio synthesized Ag NPs from *M. anisopliae*, *B. bassiana*, Bio Magic, Bio Power and Bio Catch, the mortality rates recorded were 87, 87, 80, 76, and 64 %, respectively. *M. anisopliae* was the highest % mortality (97%) in the larvae of *C. vittata,* when treated with the Nano-particles or with the fungal spores after 6 days, while the lowest (64%) was when treated with Bio Catch, *V. lecanii*. [64]

Table (26): % Mortality of EPF on the larvae of *C. vittata* using spore suspension and bio synthesized silver nanoparticles [64]

EPF	Treated with	
	Fungal spores (Mean ± S.E)	Bio synthesized Ag NPs (Mean ± S.E)
M. anisopliae	85.0 ± 2.17	97.0 ± 1.11
B. bassian	78.0 ± 2.42	87.0 ± 2.22
Bio Magic	72.0 ± 1.13	80.0 ± 1.11
Bio Power	67.0 ± 1.22	76.0 ± 1.11
Bio Catch	58.0 ± 1.22	64.0 ± 1.10
Control	6.0	6.0
S.E (m)	1.63	1.33

Data of adult mortality of *C. vittata,* treated with *M. anisopliae*, *B. bassiana*, Bio Magic, Bio Power and Bio Catch as fungal Spores and Silver Nano-particles was presented in table (27). Seven days post treatment, the parentage of mortality of *C. vittata,* adults reached 80%.

The % mortality rates recorded 69, 65, 62, 58, and 55 % by infection with fungal spores from *M. anisopliae*, *B. bassiana*, Bio Magic, *Bio* Power and Bio Catch, respectively. While, by infection with bio synthesized Ag NPs from *M. anisopliae*, *B. bassiana*, Bio Magic, Bio Power and Bio Catch, the % mortality recorded were 80, 77, 69, 65, and 60 %, respectively.

M. anisopliae was the highest % mortality (80%) in the adults of *C. vittata,* when treated with the Nano-particles or with fungal spores after 6 days, while the lowest was (60%), when they were treated by Bio Catch.[64]

Table (27): % Mortality of EPF on the adults of *C. vittata* using spore suspension and bio synthesized silver nanoparticles [64]

EPF	Treated with	
	Fungal spores (Mean ± S.E)	Bio synthesized Ag NPs (Mean ± S.E)
M. anisopliae	69.0 ± 2.11	80.0 ± 1.22
B. bassiana	65.0 ± 1.20	77.0 ± 1.00
Bio Magic	62.0 ± 2.11	69.0 ± 1.10
Bio Power	58.0 ± 1.12	65.0 ± 3.20
Bio Catch	55.0 ± 1.11	60.0 ± 1.20
Control	6.0	7.0
S.E(m)	1.53	1.54

SILVER NANOPARTICICLES FROM ENTOMOPATHOGENIC FUNGI AGAINST THE SPINY BOLLWORM, *EARIAS INSULANA* IN MAIZE CROP

Data of the treated nymphs of *E. insulana* with *B. bassiana*, *M. anisopliae* and Bio Catch (*V. lecanii*) as fungal spores and their Silver NPs particles were presented in table (28). [65]

It appears that both treatments, fungal spores and bio-synthesized Ag NPs, have a positive impact on the EPF (Entomopathogenic Fungi) being studied, as compared to the control group.Six days post-treatment, up-to (70%) mortality rate of *E. insulana* was recorded in the treated larvae. The percent mortality rates attained 60, 50, and 47% by infection with fungal spores from *B. bassiana*, *M. anisopliae,* and *V. lecanii*, respectively. [65]

Furthermore, the percentage mortality rates were 70, 60, and 58% by infection with biosynthesized Ag NPs from *B. bassiana*, *M. anisopliae* and *V. lecanii* respectively, at the same time. *B. bassiana* recorded the highest mortality (70%) in the larvae of *E. insulana*, when treated with nanoparticles or with fungal spores after 6 days will the lowest was (47%) when treated with *V. lecanii* fungal spores. The control group, which did not undergo any treatment, maintained a constant value of 5.0. This suggests that there was no significant impact on the EPF's activity in the absence of treatment. [65]

The standard error values (S.E) provided in the last row of the table indicate the variability or margin of error associated with the mean values. The S.E for fungal spores treatment is 1.3, while for bio-synthesized Ag NPs treatment, it is 1.40. These values represent the range within which the true mean values are expected to fall. [65]

Table (28): Percent mortality of EPF on the larvae of *E. insulana,* using spore suspension and biosynthesized silver nanoparticles.[65]

EPF	Treated with	
	Fungal spores (Mean ± S.E)	Bio synthesized Ag NPs (Mean ± S.E)
B. bassiana	60.0 ± 2.3	70.0 ± 2.2
M. anisopliae	50.0 ± 1.20	60.0 ± 1.2
V. lecanii	47.0 ± 1.1	58.0 ± 1.1
Control	5.0	5.0
S.E (m)	1.3	1.40

The pupae percent of mortality of *E. insulana,* treated with *B. bassiana, M. anisopliae* and *V. lecanii* as fungal Spores and Silver nanoparticles was presented in table (2). Sex days post-treatment, mortality percent of *E. insulana* pupae was recorded: 57, 45, and 40 % by infection with fungal spores from *B. bassiana, M. anisopliae* and *V. lecanii,* respectively. [65]

Will, at the infection with biosynthesized Ag NPs from *B. bassiana, M. anisopliae* and *V. lecanii* the mortality rates recorded 65, 60, and 50%, respectively. *B. bassiana* was the highest % mortality (65%) in the pupae of *E. insulana,* when treated with the nanoparticles or with the fungal spores (57%) after 6 days, while the lowest (50%) was when treated with *V. lecanii* biosynthesized. The control group, as before, shows a constant value of 5.0, indicating no significant change or treatment effect. The standard error values (S.E) provided in the last row of the table indicate the variability or margin of error associated with the mean values. The S.E for the fungal spores treatment is 1.31, while for the bio-synthesized Ag NPs treatment, it is 1.51. These values represent the range within which the true mean values are expected to fall. [65]

Table (29): Percent mortality of EPF on the pupae of *E. insulana* using spore suspension and biosynthesized silver nanoparticles[65]

EPF	Treated with	
	Fungal spores (Mean ± S.E)	Bio synthesized Ag NPs (Mean ± S.E)
B. bassiana	57.0 ± 1.2	65.0 ± 1.11
M. anisopliae	45.0 ± 1.2	60.0 ± 1.11
V. lecanii	40.0 ± 1.2	50.0 ± 1.10
Control	5.0	5.0
S.E (m)	1.31	1.51

CONCLUSION

Entomopathogenic fungi are a specialized group of microbial organisms with traits and modes of action that render them effective as biopesticides. Hypocreales species are ubiquitous parts of the soil microbiota. Soil is important for protection against environmental extremes, for water availability, and, also as a source of insect hosts.

While the chemical pesticides are becoming obsolete, the urge for research and development of biopesticides is growing. While only a few species of *Beauveria*, *Metarhizium*, *Lecanicillium* and *Isaria* are commercially available, the recent surge in studies exploring the endophytical attributes of fungi paves the way to further research into their complex ecology.

Acknowledging their ecology better, biopesticides can potentially develop to more efficient and more diversified applications. Better grasp of their compatibility with other microbial agents, natural arthropod enemies, or plant and/or chemical compounds could generate a richer and stronger pool of available tools for pest control. Also, research should be focused on discovering the most appropriate strains or isolates per plant host. Differential rates of endophytism and degrees of fungal colonization make the case for better research into the compatibility of fungal agents with their hosts and the persistence inside the plant. Inoculation method has been shown to influence the success of colonization.

Inundative EPF releases as biological control agents against insects, mites and ticks is based on the action of the agent and not on successive generations of the fungi. Attention has predominantly focused on the technical logistics of mass production and formulation of fungal biopesticides. Mass production is concerned with creating products which are relatively cost efficient, consistently effective in the field and with a relatively long shelf life.

The Ag NPs synthesized through application of spore suspension and bio synthesized silver Nano-particles showed an efficacy against *R. ferrugineus* different stages. *M. anisopliae* had the highest potential and was more effective than all the others.

The bio control efficiency of Ag NPs synthesized by five isolates of *M. anisopliae*, *B. bassiana*, Bio Magic, Bio Power and Bio Catch on *R. ferrugineus* was proven to be effective through bio assay by spore suspension and bio synthesized silver nanoparticles. *M. anisopliae* had the highest efficiency on *R. ferrugineus* and was more effective than *B. bassiana*, Bio Magic, Bio Power and Bio Catch.

The use of the Ag NPs synthesized through the application of spore suspension and biosynthesized silver nanoparticles showed efficacy against *E. insulana* larvae and pupae stages. *B. bassiana* had the highest potential as well as it was more effective than *M. anisopliae* and *V. lecanii.*

References

1. Ainsworth, G.C. 1956 . Agostino Bassi, 1773-1856, Nature. 177: 255-257.
2. Hajek AE, St. Leger RJ. 1994. Interactions Between Fungal Pathogens and Insect Hosts. Annu Rev Entomol 39, 293- 322.
3. Inglis GD, Goettel MS, Butt TM, Strasser H. 2001. Use of hyphomycetous fungi for managing insect pests. In: Butt TM, Jackson C, Magan N (eds), Fungi As BiocontrolAgents: Progress, Problems and Potential. CABI Publishing, Wallingford, pp. 23-69.
4. Shah PA, Pell JK. 2003. Entomopathogenic fungi as biological control agents. Appl. Microbiol. Biotech 61, 413-423.
5. Medo, J. and Cagan, L. 2011. Factors affecting the occurrence of entomopathogenic fungi in soils of Slovakia as revealed using two methods. *Biological Control* 59, 200-208.
6. Chandler, D., Hay, D., Reid, A.P., 1997. Sampling and occurrence of entomopathogenic fungi and nematodes in UK soils. Appl. Soil Ecol. 5, 133–141.
7. Bidochka, M.J., Kasperski, J.E., Wild, G.A.M., 1998. Occurrence of the entomopathogenic fungi *Metarhizium anisopliae* and *Beauveria bassiana* in soils from temperate and near- Northern habitats. Can. J. Bot. 76: 1198–1204.
8. Ali-Shtayeh, M.S., Mara, A.B.B.M., Jamous, R.M., 2002. Distribution, occurrence and characterization of entomopathogenic fungi in agricultural soil in the Palestinian area. Mycopathologia 156, 235–244.

9. Klingen, I., Eilenberg, J., Meadow, R., 2002. Effects of farming system, field margins and bait insect on the occurrence of insect pathogenic fungi in soils. Agric. Ecosyst. Environ. 91, 191–198.
10. Keller SP, Kessler P, Schweizer C. 2003. Distribution of insect pathogenic soil fungi in Switzerland with special reference to *Beauveria brongniartii* and *Metharhizium anisopliae*. Biocontrol 48, 307-319.
11. Meyling NV and Eilenberg J. 2006. Isolation and characterization of Beauveria bassiana isolates from phylloplanes of hedgerow vegetation. Mycol. Res. 110, 188-195.
12. Abdel-Raheem, M. A. 2005. Possibility of Using the Entomopathogenic Fungi *Beauveria bassiana* and *Metarhizium anisopliae* for controlling the Sugar-Beet Insects *Cassida vittata* Vill. and *Scrobipalpa ocellatella* Boh. in Egypt, Ph.D. Faculty of Agriculture, Cairo University, Cairo.86Pp.
13. Abdo C, Nemer N, Nemer G, Abou Jawdah Y, Kawar N. 2008. Isolation of *Beauveria* species from Lebanon and evaluationof its efficacy against the cedar web-spinning sawfly, *Cephalciata annourinensis.* Biocontrol 53, 341-352.
14. Brownbridge M, Reay SD, Cummings NJ. 2010. Association of entomopathogenic fungi with exotic bark beetles in New Zealand pine plantations. Mycopathologia, 169(1): 75-80.
15. Glare TR, Reay SD, Nelson TL, Moore R. 2008. *Beauveria caledonica* is a naturally occurring pathogen of forest beetles. Mycol. Res. 112(3), 352-360.
16. Santoro PH, Neves PM, Alexandre TM, Sartori D, Alves LF, Fungaro MH. 2008. Selection of *Beauveria bassiana* isolates to control *Alphitobi diaperinus*. J. Invertebr. Pathol. 97(2), 83-90.

17. Vu VH, Hong SI, Kim K. 2007. Selection of entomopathogenic fungi for aphid control. J. Biosci. Bioeng. 104(6), 498- 505.
18. Meyling, N.V., Eilenberg, J., 2007. Ecology of the entomopathogenic fungi *Beauveria bassiana* and *Metarhizium anisopliae* in temperate agroecosystems: potential for conservation biological control. *Biological Control* 43, 145–155.
19. Veen KH, Ferron P. 1966. A selective medium for isolation of *Beauveria tenella* and of *Metarrhizium anisopliae*. J. Invertebr. Pathol. 8, 268-269.
20. Shimazu M, Sato H. 1996. Media for selective isolation of an entomogenous fungus, *Beauveria bassiana* (Deuteromycotina: Hyphomycetes). Appl. Entomol. Zool. 31, 291-298.
21. Zimmermann G. 1998. Suggestion for a standardized method for reisolation of entomopathogenic fungi from soil using the bait method. "Insect pathogens and insect parasitic nematodes". IOBC Bull 21(4), 289.
22. Zimmermann, G. 1986. The *Galleria* bait method for detection of entomopathogenic fungi in soil. J. Appl. Entomol. 102(2) 213-215.
23. Soper, R.S., Shimazu, M., Humber, R.A. and Hajek, A.E. 1988. Isolation and characterization of *Entomophaga maimaiga* sp. nov., a fungal pathogen of gypsy moth *Lymantria dispar,* from Japan. J. Invertebr. Pathol. 51, 229-241.
24. Humber, R.A. 1996. Fungal pathogen of the Chrysomelidae and prospects for their use in biological control. In: Biology of Chrysomilidae (eds P.H. Jolivet, M.L. Cox & T.H. Hsiao) Vol. IV, SPB Academic Publishing, The Hague.
25. Zimmermann, G. 1986. The *Galleria* bait method for detection of entomopathogenic fungi in soil. J. Appl. Entomol. 102(2) 213-215.

26. Saleh, M.M.E., M.A. Abdel-Raheem., I.M. Ebadah and Huda H. Elbehery. 2016. Natural Abundance of Entomopathogenic Fungi in Fruit Orchards and their Virulence against *Galleria mellonella* larvae. Egyptian Journal of Biological Pest Control, 26(2):203-207.

27. Lacey, L.A. 1997. Manual of Techniques in Insect Pathology. Academic Press (Elsevier), NY. 409 pp.

28. Humber, R.A. 1996. Fungal pathogen of the Chrysomelidae and prospects for their use in biological control. In: Biology of Chrysomilidae (eds P.H. Jolivet, M.L. Cox & T.H. Hsiao) Vol. IV, SPB Academic Publishing, The Hague.

29. Salem, S.A.; Abdel-Raheem, M.A.; Abdel-Salam,A.M.A. and Farage,N.A. 2015. Lab-Field Evaluation of Some Egyptian isolates of Entomopathogenic Fungi *Metarhizium anisopliae* and *Beauveria bassiana* Against Sugar Beet Beetle *Cassida vittata* Vill. Swift journals of Agricultural Research, 1(2):9-14.

30. Abdel-Raheem, M.A.; I.A. Ismail; R.S. Abdel-Rahman; N.A. Farag and I. E., Abdel-Rhman.2015. Entomopathogenic fungi, *Beauveria bassiana* (Bals.) and *Metarhizium anisopliae* (Metsch.) As biological Control Agents on Some Stored Product Insects. Journal of Entomology and Zoology Studies, 3(6)316-320.

31. Abdel-Raheem, M. A.; Naglaa F. Reyad ;Abdel-Rahman, I. E.and Al-Shuraym, Laila. A. 2016. Evaluation of some isolates of entomopathogenic fungi on some insect pests infesting potato Crop in Egypt. International Journal of ChemTech Research, 9(8):479-485.

32. Roberts, D. W. 1980. Fungi toxins. In "Microbial control of Insects, Mites, and Plant Diseases" (H.D.Bureges, ed.) Vol., 2: 223-248 Academic Press, New York.
33. Clarkson, J. M. and A. K. Charnley .1996. "New insights into the mechanisms of fungal pathogenesis in insects." Trends Microbiol 4(5): 197-203.
34. Boucias, D. G., J. C. Pendland, et al. 1988. "Nonspecific factors involved in the attachment of entomopathogenic deuteromycetes to host insect cuticle." Appl. Environ. Microbiol. 54(7): 1795-1805.
35. Latge, J. P., M. Monsigny, et al. 1988. "Visualization of exocellular lectins in the entomopathogenic fungus *Conidiobolus obscurus*." J. Histochem.Cytochem. 36: 1419- 1424.
36. St Leger, R. J., T. M. Butt, et al. 1989. "Production in vitro of appressoria by the entomopathogenic fungus *Metarhizium anisopliae*." Exp. Mycol. 13: 274-288.
37. Hackman, R. H. 1984. Cuticle: Biochemistry. Biology of the Integument. J. Bereiter-Hahn, A. G. Matolts and K. S. Richards. Berlin, Springer-verlag: 626-637.
38. Neville, A. C. 1984. Cuticle: Organisation. Biology of the integument. J. Bereiter-Hahn, A. G. Matolts and K. S. Richards. Berlin, Springer-Verlag: 611-625.
39. St Leger, R. J., T. M. Butt, et al. 1989. "Synthesis of proteins including a cuticle-degrading protease during differentiation of the entomopathogenic fungus *Metarhizium anisopliae*." Exp. Mycol. 13: 253-262.
40. Kmitowa K; Bajan C, 1982. Pathogenicity level of various strains of *Beauveria bassiana* (Bals.) Vuill. Polish Ecological Studies, 8(3/4):409-417

41. Bidochka MJ; Miranpuri GS; Khachatourians GG, 1993. Pathogenicity of *Beauveria bassiana* (Balsamo) Vuillemin toward lygus bug (Hem., Miridae). Journal of Applied Entomology, 115(3):313-317.
42. Hegedus, D. D., T. A. Pfeifer, et al. 1991. "Cloning and analysis of five mitochondrial tRNAencoding genes from the fungus *Beauveria bassiana.*" Gene 109(1): 149-54.
43. Goettel, M. S., R. J. St Leger.1990. "Pathogenicity and growth of *Metarhizium anisopliae* stably transformed to benomyl resistance." Curr. Genet. 17: 129-132.
44. Florez, F. J. P. 2002. Fungi for coffee berry borer control- Colombia. 35th annual meeting of the Society of Invertebrate Pathology, Foz Do Iguassu, Brazil.
45. Kmitowa K; Kabacik Wasylik D; Wasylik D Kabacik, 1971. An attempt at determining the pathogenicity of two species of entomopathogenic fungi in relation to Carabidae. Ekologia Polska, 19(43):727-733.
46. Zaki, F.N. and Abdel-Raheem, M.A.2010. Using of Entomopathogenic fungi and insecticide against some insect pests attacking peanuts and sugar beet. Archives of phytopathology and plant protection, 43(18):1819-1828.
47. Abdel-Raheem, M. A. 2005. Possibility of Using the Entomopathogenic Fungi *Beauveria bassiana* and *Metarhizium anisopliae* for controlling the Sugar-Beet Insects *Cassida vittata* Vill. and *Scrobipalpa ocellatella* Boh. in Egypt, Ph.D. Faculty of Agriculture, Cairo University, Cairo, 86Pp.
48. Abdel-Raheem, M. A.; Naglaa F. Reyad; Abdel-Rahman, I. E.and Al-Shuraym, Laila. A. 2016. Evaluation of some isolates of entomopathogenic fungi on some insect pests infesting potato Crop in Egypt. International Journal of ChemTech Research, 9(8):479-485.

49. Hussein K. A. , Abdel-Rahman M. A. A., Abdel-Mallek A. Y., El-Maraghy S. S., Joo J. H. 2012. Pathogenicity of *Beauveria bassiana* and *Metarhizium anisopliae* against *Galleria mellonella*. Phytoparasitica 40:117–126.
50. Gupta. S. C., Leathers T.D., El-Sayed, G. N. and Ignoffo C. M. 1994. Relationships among Enzyme Activities and Virulence Parameters in *Beauveria bassiana* Infections of *Galleria mellonella* and *Trichoplusia ni*. J. Invertebr. Pathol. 64, 13-17
51. Abdel-Raheem, M.A. ; Ismail, I.A. ; Abdel-Rahman, R.S. ; Abdel-Rhman, I. E. and Naglaa F. Reyad. 2015. Efficacy of Three Entomopathogenic Fungi on Tomato leaf miner, *Tuta absoluta* in Tomato crop in Egypt. Swift journals of Agricultural Research, 1(2):15-20.
52. Magda Sabbour, and Mohamed Abdel-Raheem.2015. Efficacy of *Beauveria brongniartii* and *Nomuraea rileyi* against the potato tuber moth, *phthorimaea operculella* (zeller). *The American Journal of Innovative Research and Applied Sciences*; 1(6):197-202.
53. Abdel-Raheem, M. A.; Naglaa F. Reyad ; Al-Shuraym, Laila. A. and Abdel-Rahman, I. E.2016. Nano Entomopathogenic Fungi As biological Control Agents on Cabbage Worm, *Pieris rapae* L. (Lepidoptera: Pieridae). Der Pharma Chemica, 8(16):93-97
54. Abdel-Raheem, M.A.; and Lamya Ahmed Al-Keridis.A. 2017. Virulence of three Entomopathogenic Fungi against White fly, *Bemisia tabaci* (Genn.) (Hemiptera: Aleyrodidae) in Tomato Crop. Journal of Entomology, 14(4):155-159.
55. Abdel-Raheem, M.A.; I.A. Ismail; R.S. Abdel-Rahman; N.A. Farag and I. E., Abdel-Rhman.2015. Entomopathogenic fungi, *Beauveria bassiana*

(Bals.) and *Metarhizium anisopliae* (Metsch.) As biological Control Agents on Some Stored Product Insects. Journal of Entomology and Zoology Studies, 3(6)316-320

56. Magda Sabbour and Mohamed Abdel-Raheem.2015. Toxicity of the fungus *Beauveria bassiana* and three oils extract against *Sitophilus granaries* under laboratory and store condition. American Journal of Innovative Research and Applied Sciences; 1(7):251-256.

57. Abdel-Raheem, M.A.; I.A. Ismail; N.A. Farag, R.S. Abdel-Rahman and H. H. Elbehery. 2016. Isolates, Virulence of two Entomopathogenic Fungi as biological control agent on sugar beet fly, *Pegomyia mixta* in Egypt. Der Pharma Chemica,8(18):132-138.

58. Abdel-Raheem, M. A.; I.A. Ismail, R.S. Abdel-Rahman and Wafaa M. M. EL-Baradey. 2017. Virulence of some Entomopathogenic fungi against the Oliver black scale insect, *Saissetia oleae* (Oliver) on olive trees in Egypt.Journal of Pharmaceutical, Chemical and Biological Sciences, 5(2):97-102.

59. Abdel-Rahman, R.S., Abdel-Raheem, M. A; I.A. Ismail and Wafaa M. M. EL-Baradey. 2017. The strategy of anti-soft scale insect *Pulvinaria tenuivalvata* (Newstead) Infesting Sugar-cane. Journal of Pharmaceutical, Chemical and Biological Sciences, 5(2): 125-132.

60. Abdel-Raheem, M. A. 2005. Possibility of Using the Entomopathogenic Fungi *Beauveria bassiana* and *Metarhizium anisopliae* for controlling the Sugar-Beet Insects *Cassida vittata* Vill. and *Scrobipalpa ocellatella* Boh. in Egypt, Ph.D. Faculty of Agriculture, Cairo University, Cairo, 86Pp.

61. Abdel-Raheem, M. A. , Zakia A.Ragab and Abdel-Rahman, I. E. 2011. Effect of Entomopathogenic Fungi on the Green Stink Bug, *Nezara viridula* L. in Sugar Beet. Bull. NRC, 36(2):145-152.
62. Abdel-Raheem M. A.; Huda A. ALghamdi, Naglaa F. Reyad (2019) Virulence of Fungal Spores and Silver Nano-particles from Entomopathogenic Fungi on the Red Palm Weevil, Rhynchophorus ferrugineus Olivier (Coleoptera: Curculionidae), Egyptian Journal of Biological Pest Control, 2019, 29(1),97. https://link.springer.com/content/pdf/10.1186/s41938-019-0200-2.pdf https://doi.org/10.1186/s41938-019-0200-2
63. Abdel-Raheem M. A.; Naglaa F. Reyad, and Huda A. Alghamdi (2020) Virulence of Nano – Particle preparation of Entomopathogenic fungi and Entomopathogenic Bacteria against red palm weevil, *Rhynchophorus ferrugineus* (Olivier) (Coleoptera: Curculionidae), Romanian Biotechnological Letters,25(1):1151-1159. https://www.e-repository.org/rbl/vol.25/iss.1/5.pdf Clarivate Analytics (SCIE, BIOSIS Previews)
64. Abla F. A. Saad and Abdel-Raheem, M. A. (2020) Silver Nano-particles from Entomopathogenic Fungi on The tortoise beetle, *Cassida vittata* (Coleoptera: Chrysomelidae), Plant Archives, 20(2): 5367-5372. http://www.plantarchives.org/20-2/5367-5371%20(7053).pdf
65. Abdel-Raheem M. A.* , Ismail A. Ismail , Ragab S. Abdel-Rahman (2023) SILVER NANOPARTICICLES FROM ENTOMOPATHOGENIC FUNGI AGAINST THE SPINY BOLLWORM, EARIAS INSULANA IN MAIZE CROP, Mesopotamia Journal of Agriculture, 51(2): 98`-106.

https://magrj.mosuljournals.com/article_179646_45440b91a12390170b472c5eab999385.pdf

About Authors

Prof. Dr. Mohamed Abdel-Raheem Ali Abdel-Raheem

Professor of Entomology (Biological Control)

Pests & Plant Protection Department

Agricultural and Biological Research Institute

National Research Centre.

Prof. Dr. Mohamed Abdel-Raheem Ali Abdel-Raheem, Prof. of Entomology (Biological Control), Pests & Plant Protection Department, Agricultural and Biological Research Institute, National Research Centre, Cairo, Egypt. Published publications (307) papers (82), Books and Book chapters (225), (Scopus) h-index (11), Citations (304), (Google Scholar) h-index (19) , Citations (1073), (Web of Science) Publications (9), h-index (5), Citations (51), Research Gate, h-index (14), Citations (646), Reviewers in International Journals (143), Reviewed articles in Biological Control (817), Editor in Chief in Journal (7), Assistant Editor (1), Associated Editor in Chief in Journal (5), Editorial board in Journals (36), Advisory board (1), conference Attended (27), workshop Attended (296), Symposium Attended (178), Forum Attended (4), Prize (4), Attended Others (30), Projects As PI and member (19), Training courses For Agric. Engineering (61), Training courses For Students of University (8), attending Trainees courses (19), TV, Radio & Newspaper (26), Supervisor on Ph.D. thesis (1), Committee Ph.D. thesis (2), and member in scientific Foundation (14), Giving International lectures (4).

National Research Centre. 33rd ElBohouth St., Dokki, Cairo, Egypt.
Email Address: abdelraheem_nrc@hotmail.com, ma.abdel-raheem@nrc.sci.eg, abdelraheem_nrc@yahoo.com, profmohamedabdelraheem@gmail.com, abdelraheem196942@gmail.com, Mohamedabdelraheem42@gmail.com, Mabdelraheem69@gmail.com, mohamed196942@yahoo.com, mabdel-raheem@hotmil.com
Mobile Phone: (+2) 01155527583 – (+2) 01009580797
Scopus Author ID : 55220661700
https://www.scopus.com/authid/detail.uri?authorId=55220661700
http://scholar.google.com/citations?hl=en&user=gBtssEgAAAAJ

https://orcid.org/my-orcid?orcid=0000-0001-9240-064X
https://www.webofscience.com/wos/author/searchAbdel-Raheem, Mohamed Abdel-Raheem Ali - Web of Science Core Collection
https://www.researchgate.net/profile/Mohamed_Abdel-Raheem4/publications?sorting=recentlyAdded
http://livedna.org/20.16018
http://www.resarcherid.com/rid/R-6264-2017
https://www.webofscience.com/wos/author/record/R-6264-2017
https://www.facebook.com/share/p/1Piu6xyFmN/

- **TV, Radio & Newspaper**

1. https://www.youtube.com/attribution_link?a=NsH2xGv9mY8&u=%2Fwatch%3Fv%3Dsb46rBG8-Mc%26feature%3Dshare
2. https://youtu.be/Ub8djW0tggo
3. https://m.facebook.com/story.php?story_fbid=492907069288338&id=100076011247165
4. https://youtu.be/dOujUAZmTdw
5. https://m.facebook.com/story.php?story_fbid=492907069288338&id=100076011247165
6. https://fb.watch/gPGrDuzZFX/
7. https://www.youtube.com/watch?si=4UfyQaeHVLPiVR5J&v=nXV4NT-no7o&feature=youtu.be
8. https://www.youtube.com/watch?v=sRH44W6TbcQ
9. https://youtu.be/uq9Qj3AEBHk?si=ymaX06huG2AXJESu
10. https://www.facebook.com/share/v/Qg5s6HW3hD8PLBzp/?mibextid=xfxF2i
11. https://el-yom.com/news/archives/1092767

12. https://el-yomnews.com/?p=1156515
13. https://el-yomnews.com/%D8%A7%D9%84%D8%AF%D9%83%D8%AA%D9%88%D8%B1-%D8%B9%D8%A8%D8%AF%D8%A7%D9%84%D8%B1%D8%AD%D9%8A%D9%85-%D9%85%D8%AD%D9%85%D8%AF-%D8%B9%D9%84%D9%8A-%D9%8A%D8%AA%D8%AD%D8%AF%D8%AB-%D9%84-%D9%84%D9%84/
14. https://el-yomnews.com/?p=1179515
15. https://www.facebook.com/share/v/4MM85wKRPxsYLqTr/
16. http://https//el-yomnews.com/%D8%A7%D9%84%D8%AD%D8%B4%D8%B1%D8%A7%D8%AA-%D9%81%D9%8A-%D8%AE%D8%AF%D9%85%D8%A9-%D8%A7%D9%84%D8%B9%D8%AF%D8%A7%D9%84%D8%A9-%D9%85%D8%AD%D8%A7%D8%B6%D8%B1%D8%A9-%D8%AA%D9%83%D8%B4%D9%81-%D8%AF%D9%88/
17. https://youtube.com/watch?v=F21ziQxdeco&si=LvSUixmnSg4JM1zK
18. https://youtu.be/F21ziQxdeco?si=jyAwbU8dVwkYzRH4
19- https://el-yomnews.com/المكافحة-الحيوية-بالفطريات-حلول-مبتك/

Eng. Mai Magdy Mousa

Shebin El-Kom, Menoufia, Egypt

Kolibri 360 GmbH
Ferdinand-Jühlke-Straße 7
99095 Erfurt, Deutschland
produktsicherheit@kolibri360.de

Druck:
CPI Druckdienstleistungen GmbH
im Auftrag der
Kolibri 360 GmbH
Ferdinand-Jühlke-Str. 7
99095 Erfurt